普通高等院校专业英语系列教材

上海市高水平地方应用型高校建设资金资助

基础电力英语导读

主编◎谢华

副主编◎周玉珍 李亚萍 胡汝昉

AN ENGLISH GUIDE

TO ELECTRIC POWER

华中科技大学出版社
http://www.hustp.com
中国·武汉

内 容 提 要

　　本书主要涉及能源与电力行业的基础知识,全书包含10个单元,主要介绍了能源分类的基本形式,我国电力工业的基本状况,燃煤火力发电机组、先进燃气轮机联合循环发电机组等的基本生产过程及设备状况,电学及其基本发展历程,电力系统和电气设备的基本参数、设备和工作特点,安全用电的基本常识,太阳能发电、风力发电的基本形式和基础工作原理,智能电网,能源与政策,能源与技术等相关内容。

　　本书可作为高等院校能源和电力专业的通用英语类教学用书,帮助学习者了解能源和电力的基础知识,也可作为研究生、本科生、大专生及电力行业相关从业人员学习专业英语的辅导用书。

图书在版编目(CIP)数据

基础电力英语导读/谢华主编. —武汉:华中科技大学出版社,2020.10
ISBN 978-7-5680-6703-4

Ⅰ. ① 基… Ⅱ. ① 谢… Ⅲ. ① 电力工业—英语—教材 Ⅳ. ① TM

中国版本图书馆 CIP 数据核字(2020)第 196723 号

基础电力英语导读　　　　　　　　　　　　　　　　　　　　　　　　谢　华　主编
Jichu Dianli Yingyu Daodu

策划编辑:宋　焱
责任编辑:王青青
封面设计:廖亚萍
责任校对:张会军
责任监印:周治超
出版发行:华中科技大学出版社(中国·武汉)　　　　电话:(027)81321913
　　　　　武汉市东湖新技术开发区华工科技园　　　　邮编:430223
录　　排:华中科技大学出版社美编室
印　　刷:武汉科源印刷设计有限公司
开　　本:787mm×1092mm　1/16
印　　张:14　　插页:1
字　　数:320 千字
版　　次:2020 年 10 月第 1 版第 1 次印刷
定　　价:49.00 元

前言
Preface

  本书是编者在阅读大量专业文献的基础上,悉心设计、精选精编而成。其特点是内容丰富、覆盖面广,既有专业知识的英文描述,又有扩充的实际例证。本书不仅以服务本科教学为主要目标,向学生普及能源电力基础知识,激发学生对能源电力知识的兴趣,帮助其了解我国未来能源发展趋势,也力求使研究生、大专生及电力行业的相关从业人员从中获益。

  本书涉及能源与电力行业的基础知识,全书包含 10 个单元,主要介绍了能源分类的基本形式,我国电力工业的基本状况,不同发电机组的基本生产过程及设备状况,电学及其基本发展历程,电力系统和电气设备的基本参数、设备和工作特点,安全用电的基本常识,新能源发电的基本形式和基础工作原理,智能电网,能源与政策,能源与技术等相关内容。书中每个单元包含主要知识讲解和扩展阅读,正文后附有生词表和形式多样的练习题,供读者检验自己对本单元内容的掌握程度,以便巩固提高。

  本书引用图文的版权、著作权等均属于原作品版权人、著作权人所有。本书创作团队衷心感谢所有原作品的相关版权权益人及所属公司对外语本科教育的大力支持。本书在编著过程中使用的部分图文,由于客观原因,我们无法与相应作者取得联系,如书中内容侵犯到您的著作者权益,请联系我们。

  本书涉及的内容较广,由于编者水平有限和时间紧迫,书中难免有疏漏之处,敬请读者批评指正。

编　者

2020 年 3 月

目录
Contents

Unit 1

Introduction to Energy

In today's society, electricity exists in every line of work, and everyone can not live without electricity. Living organisms require energy to stay alive, such as the energy humans get from food. Human civilization requires energy to function, which it gets from energy resources such as fossil fuels, nuclear fuels, or renewable energy. The processes of the earth's climate and ecosystem are driven by the radiant energy the earth receives from the sun and the geothermal energy contained within the earth. Therefore, understanding energy power and making good use of energy power (see Fig. 1.1) have become an indispensable knowledge and ability of modern life and work.

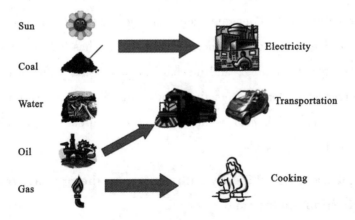

Figure 1.1 Uses of energy sources

Source: https://www.slideshare.net/ElviraParsCortacans/the-energy-year-5.

1. The history of human energy use

The history of human evolution is also a history of continuously asking for energy from nature. With the development and utilization of energy, human society has gradually shifted from ancient slash-and-burn cultivation to modern civilization.

Four historical periods of human energy use

1) Firewood period: from ancient times to the middle of the 18th century

About 500,000 years ago, humans learned to use tools and fire. The fuel at this time was mainly branches, weeds, and etc. Humans completed the evolution from apes to humans, and energy use entered the period of firewood. Human beings used firewood for cooking, heating and lighting. They were powered by manpower, animal power, and a small amount of watermills and windmills. They were engaged in manual production and transportation. Productivity was low during this period, and mankind was constrained by nature.

2) Coal period: from the middle of the 18th century to the beginning of the 20th century, represented by the invention of the steam engine

Beginning in the 17th century, the British began to mine and use coal on a large scale. The use of gas lamps ended the long night of mankind. In the middle of the 18th century, the invention of steam engine made coal become the second generation of main energy. The wide application of coal-fueled steam engines has led to rapid development of textile, metallurgical, mining, and mechanical processing industries. At the same time, the emergence of steam locomotives and ships has made great progress in the transportation industry. With coal as the main power source, mankind has begun to transform nature on a large scale.

3) The oil era: from the early 20th century to the present, represented by internal combustion engines and electricity

The Chinese have long discovered that petroleum is a flammable liquid, and Shen Kuo of the Song Dynasty made a detailed record of petroleum. In 1854, the state of Pennsylvania launched the world's first oil well, and the modern oil industry began. In

1886, the Germans Benz and Daimler developed the world's first car powered by gasoline and driven by an internal combustion engine, and began the era of automobiles using petroleum. At the beginning of the 19th century, Faraday discovered the phenomenon of electromagnetic induction. At the end of the 19th century, Edison invented the electric lamp. As a primary energy source, oil and coal were converted into a secondary energy source — electricity that was more convenient to transport and use. In the 1960s, global oil consumption surpassed coal and oil became the third generation of energy. Based on petroleum and electrical energy, the invention and use of automobiles, ships, aircraft, electric locomotives, power stations, and information equipment such as telephones, televisions, and computers have pushed humankind to the modern era of civilization.

4) Renewable energy period: from the middle of the 21st century, the combined use of multiple energy sources

In the 21st century, mankind will gradually enter the era of renewable energy. Traditional fossil energy sources are facing depletion and environmental pollution. Fossil energy sources such as oil, coal, and natural gas will gradually be launched on the stage of history.

Hydropower, as well as renewable energy sources such as solar, wind, ocean, and biomass, has gradually taken the stage of history and played a leading role. According to the forecast of the current development trend, by the middle of the 21st century, energy use is likely to be a combination of multiple energy sources including hydropower, solar energy, wind energy, biomass energy and other renewable energy sources, as well as nuclear energy (nuclear fission and fusion).

2. The definition and classification of energy

2.1　The definition of energy

What is energy? Broadly speaking, energy is a natural resource that can provide energy to humans. Fossil energy sources such as coal and petroleum provide heat, water and wind force provide mechanical energy, geothermal provides thermal energy,

and the sun provides electromagnetic radiation (which can be converted into heat or electricity). Many natural processes produce a certain amount of energy, and even ordinary garbage can also produce energy, but the number of transformations and the ease of transformation are very different.

Therefore, energy can be defined as a more concentrated and easily converted energetic substance or energy resource.

2.2 The classification of energy

There is a wide variety of energy, and through the continuous development and research of human beings, more new energy sources have begun to meet human needs. Depending on the way of division, energy can also be divided into different types. There are eight main divisions.

1) Based on source of energy

Energy can be divided into three main categories by source. (1) Energy from the sun. These include energy directly from the sun (e. g. solar thermal radiation) and indirect energy from the sun (e. g. coal, oil, natural gas, oil shale and biomass such as fuelwood, hydro and wind). (2) Energy from the earth itself. One is geothermal energy, which exists in the earth's interior, such as underground hot water, underground steam and hot dry rock, and the other is the atomic nuclear energy contained in nuclear fuel such as uranium and thorium in the earth's crust. (3) The energy generated by objects such as the moon and the sun to the earth's gravity, such as tidal energy.

2) Based on form of energy

According to the basic form of energy, energy can be divided into primary energy and secondary energy (see Tab. 1.1). The former is natural energy, referring to the existing energy in nature, such as coal, oil, natural gas, hydropower and so on. The secondary energy is different from the primary energy. It is not directly taken from the natural world and can only be obtained after the primary energy processing and conversion. It refers to energy products, such as electricity, gas, steam and petroleum products, converted from primary energy processing. The secondary energy can also be interpreted as the energy that is reused from the primary energy. For example, the steam generated by burning coal to drive a generator can be called secondary energy. Or after the electric energy is used, it is converted into wind energy by electric fans. At this time, wind energy can also be called secondary energy. There must be a certain degree of loss between the secondary energy and the primary energy.

Table 1.1　The classification of energy

Primary energy	Renewable energy	direct solar radiation, biomass, wind, flowing water, waves, tides, ocean currents, sea water temperature differences, geothermal water, geothermal steam, hot rock formations
	Non-renewable energy	fossil fuels (coal, petroleum, natural gas, oil shale, combustible ice), nuclear fuels (uranium, plutonium, deuterium, etc.)
Secondary energy	electricity, hydrogen, gasoline, kerosene, diesel, gunpowder, alcohol, methanol, propane, aniline, coke, nitrocellulose, nitroglycerin, etc.	

3) Based on properties of energy

There are fuel-based energy sources (coal, oil, natural gas, peat, wood) and non-fuel-based energy sources (water, wind, geothermal energy, marine energy). Humans began to use energy other than their physical strength by using fire. The earliest fuel was wood, and later they used various fossil fuels, such as coal, oil, natural gas, and peat. Solar energy, geothermal energy, wind energy, tidal energy and other new energy sources have been studied for use. Fossil fuels are currently consumed in large quantities, and the reserves of these fuels on the planet are limited. In the future, uranium and plutonium will provide most of the energy needed by the world. Once the technical problems of controlling nuclear fusion are solved, humanity will actually have endless energy.

4) Based on pollution to the environment

According to whether energy consumption causes environmental pollution, energy can be divided into clean energy and non-clean energy. Clean energy includes hydro energy, electricity, solar energy, wind energy and nuclear energy, non-clean energy includes coal, oil and so on.

5) Based on usability

Energy can be divided into conventional energy and new energy based on

usability. Conventional energy sources are technically mature and commonly used, including renewable hydro energy resources and non-renewable coal, oil and natural gas resources in primary energy.

Newly-used or under-development energy is called new energy. New energy sources are compared with conventional energy sources, including solar energy, wind energy, geothermal energy, ocean energy, biomass energy, hydrogen energy, and nuclear fuel for nuclear power generation. Because the energy density of new energy is low, or the grade is low, or intermittent, the economy of conversion and utilization according to the existing technical conditions is still not high. New energy is still in the research and development stage and can only be developed and utilized according to local conditions; but most of the new energy is renewable energy, being rich in resources and widely distributed, and it will be one of the main energy sources in the future.

6) Based on long-term availability

Primary energy can further be classified. Any energy that can be continuously replenished or can be regenerated within a short period of time is called renewable energy, and the other is called non-renewable energy. Wind, hydro, ocean, tidal, solar and biomass energy are renewable energy sources; coal, oil and natural gas are non-renewable energy sources. Geothermal energy is basically non-renewable energy, but judging from the huge reserves inside the earth, it is also renewable. New developments in nuclear energy will make the nuclear fuel cycle multiply. The energy of nuclear fusion can be 5 to 10 times higher than the energy of nuclear fission. The most suitable fuel for nuclear fusion, deuterium, exists in seawater in large quantities, which can be described as inexhaustible. Nuclear energy is one of the pillars of the future energy system.

7) Based on commercial application

Anything entering the energy market as a commodity, such as coal, oil, natural gas and electricity, is a commercial energy source, which is consumed in large quantities as a commodity through circulation. Non-commercial energy refers to the on-site use of agricultural waste such as fuelwood and straw, and human and animal waste. Non-commercial energy accounts for a large proportion of energy supply in rural areas of developing countries. This energy is not the object of a commodity transaction, so it is difficult to calculate in the energy balance. In 1975, the world's non-commercial energy was about 0. 6 terawatt-years, equivalent to 600 million tons of standard coal.

8) Based on morphological characteristics

This classification is at the level of transformation and application. The types of energy recommended by the World Energy Commission are divided into solid fuel, liquid fuel, gas fuel, hydropower, electric energy, solar energy, biomass energy, wind energy, nuclear energy, marine energy and geothermal energy. Among them, the first three types are collectively referred to as fossil fuels or fossil energy. The above-mentioned energy that has been recognized by human beings can be converted into some form of energy that people need under certain conditions. For example, when fuelwood and coal are heated to a certain temperature, they can be combined with oxygen in the air to release a large amount of heat energy. We can use heat for heating, cooking, or cooling, or we can use heat to generate steam, and use steam to propel the steam turbine to turn thermal energy into mechanical energy. We can also use steam turbines to drive generators to turn mechanical energy into electricity. It can be converted into mechanical energy, light energy or thermal energy by factories, enterprises, institutions, farming, animal husbandry and forest areas and households.

3. Energy conversion and conservation

3.1 Energy conversion

Energy conversion, also known as energy transformation, is the process of changing energy from one form to another (see Fig. 1. 2). In physics, energy is a quantity that provides the capacity to perform work (e. g. lifting an object) or provides heat. In addition to being convertible, according to the law of conservation of energy, energy is transferable to a different location or object, but it cannot be created or destroyed.

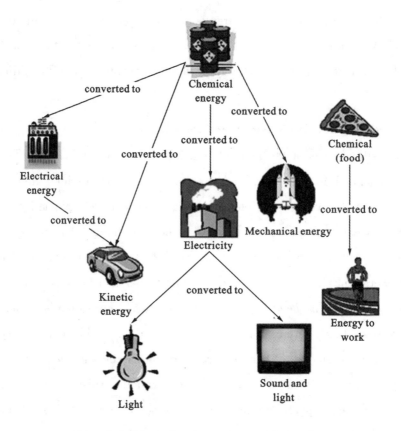

Figure 1. 2　Examples of sets of energy conversions

Source：https：//quizlet. com/118233440/energy-conversions-flash-cards/.

Many forms of energy may be used in natural processes to provide some service to society such as heating, refrigeration, lighting or performing mechanical work to operate machines. For example, to heat a home, the furnace burns fuel, whose chemical potential energy is converted into thermal energy, which is then transferred to the home's air to raise its temperature.

Conversions to thermal energy from other forms of energy may occur with 100% efficiency. Conversion among non-thermal forms of energy may occur with fairly high efficiency, though there is always some energy dissipated thermally due to friction and similar processes. Sometimes the efficiency can be close to 100% when an object is dropped into a vacuum, the potential energy is converted into kinetic energy. This also applies to the opposite case, for example, an object in an elliptical orbit around another body converts its kinetic energy (speed) into gravitational potential energy (distance from the other object) as it moves away from its parent body. When it reaches the furthest point, it will reverse the process, accelerating and converting potential energy into kinetic. Since space is a near-vacuum, this process has close to 100% efficiency.

Thermal energy is very unique because it can not be converted to other forms of energy. Only a difference in the density of thermal / heat energy (temperature) can be used to perform work, and the efficiency of this conversion will be less than 100%. This is because thermal energy represents a particularly disordered form of energy; it spreads out randomly among many available states of a collection of microscopic particles constituting the system (these combinations of position and momentum for each of the particles are said to form a phase space). The measure of this disorder or randomness is entropy, and its defining feature is that the entropy of an isolated system never decreases. One can not take a high-entropy system (like a hot substance, with a certain amount of thermal energy) and convert it into a low entropy state (like a low-temperature substance, with correspondingly lower energy), without that entropy going somewhere else (like the surrounding air). In other words, there is no way to concentrate energy without spreading out energy somewhere else.

In order to make energy transformation more efficient, it is desirable to avoid thermal conversion. For example, the efficiency of nuclear reactors, where the kinetic energy of the nuclei is first converted to thermal energy and then to electrical energy, lies at around 35%. By direct conversion of kinetic energy to electric energy, affected by eliminating the intermediate thermal energy transformation, the efficiency of the energy transformation process can be dramatically improved.

3.2　Energy conservation

Energy conservation is the effort made to reduce the consumption of energy by using less of an energy service. This can be achieved either by using energy more efficiently (using less energy for a constant service) or by reducing the amount of service used (for example, by driving less). Energy conservation is a part of the concept of eco-sufficiency. Energy conservation reduces the need for energy services and can result in increased environmental quality, national security, personal financial security and higher savings. It is at the top of the sustainable energy hierarchy. It also lowers energy costs by preventing future resource depletion.

Energy can be conserved by reducing wastage and losses, improving efficiency through technological upgrades and improved operation and maintenance. On a global

level energy use can also be reduced by the stabilization of population growth.

Energy can only be transformed from one form to another, such as heat energy to kinetic energy power in cars, or kinetic energy of water flow to electricity in hydroelectric power plants. However, machines are required to transform energy from one form to another. The wear and friction of the components of these machines while running cause losses of very high amounts of energy and very high related costs. It is possible to minimize these losses by adopting green engineering practices to improve life cycle of the components.

4. Evaluation of energy quality

Energy quality is the contrast between different forms of energy, the different trophic levels in ecological systems and the propensity of energy to convert from one form to another. The concept refers to the empirical experience of the characteristics, or qualia, of different energy forms as they flow and transform. It appeals to our common perception of the heat value, versatility, and environmental performance of different energy forms and the way a small increment in energy flow can sometimes produce a large transformation effect on both energy physical state and energy. For example, the transition from a solid state to liquid may only involve a very small addition of energy. Methods of evaluating energy quality are sometimes concerned with developing a system of ranking energy qualities in hierarchical order.

The ranking and scientific analysis of energy quality was first proposed in 1851 by William Thomson under the concept of "availability". It was later continued and standardized in Japan. I. Dincer and Y. A. Cengel (2001) stated that energy forms of different qualities are now commonly dealt with in steam power engineering industry. However, energy engineers were aware that the notion of heat quality involved the notion of value. A. Thumann wrote, "The essential quality of heat is not the amount but rather its 'value'" — which brings into play the question of teleology and wider, or ecological-scale goal functions. In an ecological context, S. E. Jorgensen and G. Bendoricchio say that energy is used as a goal function in ecological models, and express energy "with a built-in measure of quality like energy".

Energy quality can be measured in the following aspects.

1) **Flow density of energy**

Flow density of energy is the power obtained from energy in a unit volume or unit area. The flow density of solar and wind energy is small, only a few hundred watts / square meter; the energy flow density of conventional energy and nuclear energy is large. Low energy flow density is a common feature of renewable energy. Therefore, the use of renewable energy such as solar energy and wind energy requires a large receiving area.

2) **Development and investment of energy**

Fossil energy and nuclear fuel, from surveying and mining to processing and transportation, require a lot of manpower and material resources, which consume energy. Renewable energy sources such as solar energy and wind energy are provided by nature and do not require energy costs. The main cost is a one-time investment.

One-time investment in renewable energy equipment is still expensive. According to the current technology level, the price of power generation equipment such as solar energy and ocean energy is tens of thousands to hundreds of thousands of yuan per kW, and wind energy has fallen below 10,000 yuan per kW. Installations and hydropower equipment are priced at several thousand yuan per kW.

3) **Continuity of energy supply and possibilities of storage**

Various fossil fuels and nuclear fuels are relatively easy to meet these two requirements. Renewable energy sources such as solar energy and wind energy are difficult to store and continuously supply.

4) **Shipping costs and wastage**

Solar energy, wind energy, geothermal energy, and other new energy are difficult to transport; oil and natural gas can be piped from the production site to users; because coal is solid, transportation is much more difficult; if hydropower stations are far away from users, long-distance transmission losses are also large.

5) **Storage capacity**

Although China's coal and water resources are abundant, coal is concentrated in the north and water power is concentrated in the west, and transportation problems exist. China's coal reserves rank third in the world, and its output is the first in the

world (about 1.5 billion tons). Hydropower resources rank first in the world, but the development rate is less than 5% (hydropower accounts for about 20% of total power generation, and thermal power generation is 75%), while the United States, Canada, Japan and other countries have more than 40% hydropower development rates.

6) Energy grade

Low-grade energy is more difficult to be converted into electrical energy, that is, high-grade energy is more easily converted to electrical energy.

5. Energy and ecological environment

5.1 Greenhouse effect

Gases such as carbon dioxide (CO_2) and methane (CH_4) in the atmosphere allow short-wave radiation from the sun to pass, but can absorb infrared long-wave radiation emitted by the earth, and then radiate it back to the earth. Too much CO_2 and other gases are like glass in a greenhouse, blocking the normal heat loss of the earth (see Fig. 1.3). This is the greenhouse effect.

Figure 1.3 Air pollution from factories

Source: https://pixabay.com/zh/photos/pollution-factory-industry-smoke-2575166/.

Researchers have found that since the industrial revolution of 1840, the concentration of CO_2 in the atmosphere has increased by 30% and has increased from 280 ppmV to 368 ppmV, which may be the highest value in the past 420,000 years. At present, the CO_2 content is increasing by 3% every year. About 23 billion tons of CO_2 are released into the earth's atmosphere every year, and more than 700 tons of CO_2 are released per second.

If this trend continues, global temperatures will rise by 1.5-4.5 ℃ in the next 50 years, polar ice and snow will melt, seawater will heat up and expand, and most of the world's coastal areas will be flooded. Climate will also change significantly.

5.2　Air pollution

Air pollution includes harmful substances such as sulfur oxides (SO_2), nitrogen oxides (NOx), ozone (O_3), and dust. The main reasons are the heavy use of fossil fuels, large-scale burning of forests, and industrial exhaust emissions.

According to statistics, 1 million kilowatts of coal power stations can discharge 1 million tons of dust (dust, CO, etc.), 60,000 tons of sulfur dioxide, and 630 kg of benzopyrene, a strong carcinogen, forming acid rain, which not only damages forests and crops, but also pollutes water sources, animals and human beings.

5.3　Water pollution

The world is facing severe water shortages, especially in China. China has the world's first hydropower resources, but it is a poor country with a per capita water resource of 1/4 of the world average, 1/7 of the United States, and 1/5 of Russia.

With the further expansion of the population, our living space will be even narrower. Per capita arable land is one-third of the world average, ranking 126th out of 140 countries in the world.

The 30 major rivers across the country, from the Songhua River and Liaohe River in the northeast to the mother rivers of the Chinese nation — the Yellow River and the Yangtze River, to the Yangtze River and the Pearl River in the south, have all been seriously polluted.

Only 5% of industrial and domestic sewage is treated in the country, and the rest are discharged into rivers and lakes without any treatment. The eutrophication of lakes

represented by the "Three Lakes" (Dianchi Lake, Taihu Lake and Chaohu Lake) is quite serious. Sewage and drinking water sources are seriously polluted, causing great harm to human health.

5.4 Deforestation

Forests are an important part of human's natural ecological environment. They provide habitats for wildlife, conserve water, protect against wind and sand, mediate the climate, maintain soil and water, purify the atmosphere, absorb large amounts of CO_2 and emit O_2.

Tropical rainforests are called "the lungs of the earth", providing 50% of the world's fresh oxygen and offsetting the effects of the greenhouse effect. Since 1945, more than half of the rainforests in the world have been destroyed, and the current coverage is less than 10% of the earth. Most are distributed in the Brazilian Amazon basin in South America, and a few are distributed in countries such as Indonesia in Southeast Asia.

China is a large country with poor forests, and the situation of deforestation is quite serious. The forest coverage rate is only 12%, which is 1/2 of the world average. The total forest area accounts for 4% of the world, ranking 107th among 140 countries. Since 1980, China's original forest area has been reduced by 23%. From northeast to southwest, forests have been severely damaged. A large amount of deforestation in the Greater Khingan forest in the northeast has reduced the annual precipitation in Heilongjiang Province from 600 mm to 380 mm in the past. Drought and drowning often occur now.

5.5 Ozone hole

The ozone layer is located in the stratosphere about 30 kilometers from the ground. It has a strong absorption of solar ultraviolet rays, can absorb more than 99% of solar ultraviolet rays harmful to humans, and is a natural barrier for life on earth.

In 1985, a British Antarctic expedition discovered a hole in the ozone layer over Antarctica, where the O_3 content was only 40% to 50% of the normal value, and the area was equivalent to the size of the United States. Hollows were also found over the North Pole.

Scientists believe that the reduction of O_3 by 1% and the increase of ultraviolet radiation on the ground will increase the incidence of skin cancer and cataracts. It may also means that marine plankton would't be able to perform photosynthesis. Once the food chain is blocked, marine life will face extinction.

The main causes of ozone layer destruction are chlorofluorocarbons (CFCs) and methane produced by human activities, which are found in refrigerators, air conditioners, fire extinguishers, sprays and chemical production. It takes 100 years for the ozone layer to fully recover.

5.6   Waste pollution

According to statistics from the national environmental protection department, the total amount of industrial solid waste (excluding township and village enterprises) is more than 700 million tons per year, and the comprehensive utilization rate is only about 30%. Cumulative accumulation has exceeded 6.6 billion tons, and two-thirds of the cities are surrounded by garbage. More than 100 million tons of urban domestic garbage is produced every year, and it's growing at a rate of 6% to 7% per year, and the rate of harmless treatment is less than 20%. A large amount of untreated industrial waste slag and urban garbage are stored in suburban areas and other places, becoming a serious secondary pollution source.

According to the World Health Organization, three quarters of the 45 million deaths worldwide each year are due to environmental degradation. China's economy has developed rapidly in the past 20 years, but the environment has deteriorated. Therefore, each of us should be concerned about the improvement of our surroundings!

New words and phrases

arable   *adj.*   适于耕种的；可开垦的

benzopyrene   *n.*   苯并芘

carcinogen   *n.*   致癌物质

cataract   *n.*   白内障；洪流

chlorofluorocarbon   *n.*   含氯氟烃；氯氟化碳

constrain   *v.*   束缚

conventional   *adj.*   符合习俗的；传统的

convertible   *adj.*   可改变的

cumulative   *adj.*   累积的

deforestation   *n.*   采伐森林，森林开伐

degradation   *n.*   退化；降格

deuterium   *n.*   氘

dissipate   *v.*   浪费；使……消散

electromagnetic   *adj.*   电磁的

elliptical   *adj.*   椭圆的

energy conversion   能量转换；能量变换

entropy   *n.*   熵（热力学函数）

eutrophication   *n.*   富营养化

flammable   *adj.*   易燃的；可燃的

fossil fuel   化石燃料

friction   *n.*   摩擦

furnace   *n.*   火炉；熔炉

infrared   *adj.*   红外线的

kinetic   *adj.*   运动的；活跃的

locomotive   *n.*   机车；火车头

metallurgy   *n.*   冶金；冶金学

methane   *n.*   甲烷；沼气

microscopic   *adj.*   微观的；用显微镜可见的

momentum　*n.*　动量;动力

morphological　*adj.*　形态学的

nitrogen oxide　氮氧化合物

peat　*n.*　泥煤;泥炭块

petroleum　*n.*　石油

photosynthesis　*n.*　光合作用

plankton　*n.*　浮游生物（总称）

potential energy　势能

precipitation　*n.*　沉淀物

qualia　*n.*　性质;特性（quale 的复数）

radiant　*adj.*　辐射的

reserve　*n.*　蕴藏;储量

sewage　*n.*　污水;下水道;污物

slash-and-burn　*adj.*　刀耕火种的

steam engine　蒸汽机

stratosphere　*n.*　平流层

thorium　*n.*　钍

trophic　*adj.*　营养的;有关营养的

ultraviolet　*adj.*　紫外的;紫外线的

uranium　*n.*　铀

wastage　*n.*　损耗

 Exercises

I . Warming-up questions

Directions：Give brief answers to the following questions.

1. How can energy be classified into different types?

2. Why is it that the vast majority of human energy comes directly or indirectly from the sun?

3. How is energy converted to electricity? Please give some examples.

Ⅱ. Technical terms

Directions：Please give the Chinese or English equivalents of the following terms.

1. ozone hole   2. electromagnetic radiation

3. secondary energy   4. hydropower

5. tidal energy   6. storage capacity

7. 传统能源   8. 可再生能源

9. 能源转换   10. 内燃机

11. 机械能   12. 化石燃料

Ⅲ. Blank filling

Directions：Complete the following sentences by choosing words or phrases given below.

electromagnetic	radiant	convertible	dissipate	degradation

1. Most of the world's _____ energy comes from fossil fuels that are burned to produce heat that is then used as a transfer medium to mechanical or other means in order to accomplish tasks.

2. Desertification, is a form of land _____, by which fertile land becomes desert.

3. Snakes detect infrared signals through a mechanism involving _____ heating of the pit organ, rather than photochemical transduction.

4. The team's findings point toward an oscillator that would harness the spin of electrons to generate microwaves _____ waves in the frequencies used by mobile devices.

5. Higher sea surface temperatures cause low-level marine clouds to _____, allowing more of the sun's warming rays to break through, causing a further rise in ocean temperatures.

Ⅳ. Translation

Directions：Translate the following passages from English into Chinese or from Chinese into English.

Passage 1

In the stated Policies Scenario, energy demand rises by 1% per year to 2040. Low-carbon sources led by solar photovoltaic (PV), supply more than half of this growth, and natural gas, boosted by rising trade in liquefied natural gas (LNG), accounts for another third. Oil demand flattens out in the 2030s, and coal use edges lower. Some parts of the energy sector, led by electricity, undergo rapid transformations. Some countries, notably those with "net zero" aspirations, go far in reshaping all aspects of their supply and consumption. However, the momentum behind clean energy technologies is not enough to offset the effects of an expanding global economy and growing population. The rise in emissions slows, but with no peak before 2040, the world falls far short of shared sustainability goals.

Passage 2

核能技术较为成熟，可以大规模开发利用。核燃料是可弥补化石燃料匮乏、减少环境污染、实现大规模工业应用的能源。地球上已探明的核裂变燃料，即铀矿和钍矿资源，按其所含能量计算，相当于化石燃料总能量的 20 多倍。此外，地球上还存在大量核聚变燃料氘，如果人类实现了商用可控核聚变，按照目前世界能源消费水平，地球上氘、氚的储量可供人类使用百亿年。

 Additional reading

Unit 2
Thermal Power Generation

Electric power is an energy source that is powered by electricity. Invented in the 1870s, the invention and application of electricity set off the second wave of industrialization. Since the 18th century, science and technology have changed people's lives. The large-scale power system that emerged in the 20th century is one of the most important achievements in the history of human engineering science. It converts natural primary energy into electricity through mechanical energy devices, which are then transmitted, transformed and distributed to customers. In this unit, we are going to learn electric power and the traditional mode of thermal power generation.

1. Basic notions of electric power and electricity

Electric power is the rate, per unit time, at which electrical energy is transferred by an electric circuit.

Electric power, like mechanical power, is the rate of doing work, measured in watts, and represented by the letter P. The term wattage is used colloquially to mean "electric power in watts". The electric power in watts produced by an electric current (I) consisting of a charge of coulombs (Q) every seconds (t) passing through an electric potential (voltage) difference of V is

$$P = \text{work done per unit time} = \frac{VQ}{t} = VI$$

where Q is electric charge in coulombs, t is time in seconds, I is electric current in amperes, and V is electric potential or voltage in volts.

Electrical power provides a highly ordered form of energy and can be carried long distances and converted into other forms of energy such as motion, light or heat with high energy efficiency. It is thus used as a type of secondary energy.

Electric power, produced from central generating stations and distributed over an electrical transmission grid, is widely used in industrial, commercial and consumer applications. The per capita electric power consumption of a country correlates with its industrial development. Electric motors power manufacturing machinery and propel subways and railway trains. Electric lighting is the most important form of artificial light. Electrical energy is used directly in processes such as extraction of aluminum from its ores and in production of steel in electric arc furnaces. Reliable electric power is essential to telecommunications and broadcasting. Electric power is used to provide air conditioning in hot climates, and in some places electric power is an economically competitive source of energy for building space heating. Use of electric power for pumping water ranges from individual household wells to irrigation projects and energy storage projects.

A characteristic of electricity is that it is not freely available in nature in large amounts, so it must be "generated" (that is, transforming other forms of energy to electricity). Electricity generation is the process of generating electric power from sources of primary energy. For utilities in the electric power industry, it is the stage prior to its delivery to end users (transmission, distribution, etc.) or its storage (using, for example, the pumped-storage method). Electricity is mostly generated at a power station by electromechanical generators, driven by heat engines heated by combustion, geothermal power or nuclear fission. Other generators are driven by the kinetic energy of flowing water and wind. There are many other technologies that are used to generate electricity such as photovoltaic solar panels.

Electricity is most often generated at a power plant by electromechanical generators, primarily driven by heat engines fueled by combustion or nuclear fission but also by other means such as the kinetic energy of flowing water and wind. Other energy sources include solar photovoltaic and geothermal power.

Electricity generation at central power stations started in 1882, when a steam

engine driving a dynamo at Pearl Street Station produced a DC current that powered public lighting on Pearl Street, New York. The new technology was quickly adopted by many cities around the world, which adapted their gas-fueled street lights to electric power. Soon after electric lights would be used in public buildings, in businesses, and to power public transport, such as trams and trains.

The first power plants used water power or coal. Today a variety of energy sources are used, such as coal, nuclear, natural gas, hydroelectric, wind, and oil, as well as solar power, tidal power, and geothermal sources. Figure 2.1 shows the electricity production from 1980 to 2013 in the world. Total worldwide gross production of electricity in 2016 was 25,082 TWh. Sources of electricity were coal and peat 38.3%, natural gas 23.1%, hydroelectric 16.6%, nuclear power 10.4%, oil 3.7%, solar / wind / geothermal / tidal / other 5.6%, biomass and waste 2.3%.

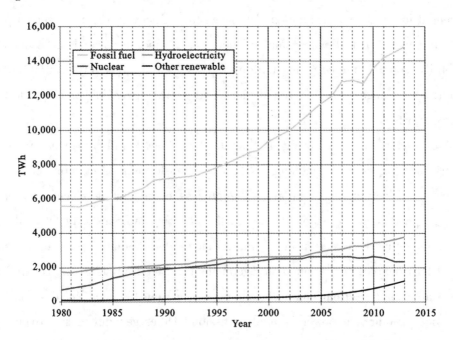

Figure 2.1 Electricity production in the world, 1980-2013

Source: electricity generation-wikipedian. wikipedia. org.

The selection of electricity generation modes and their economic viability varies in accordance with demand and region. Economic conditions vary considerably around the world, resulting in widespread residential selling prices, e. g. the price in Iceland is 5.54 cents per kWh while in some island nations it is 40 cents per kWh. Hydroelectric plants, nuclear power plants, thermal power plants and renewable sources have their own pros and cons, and selection is based upon the local power requirement and the fluctuations in demand. All power grids have varying loads on them but the daily minimum is the base load, often supplied by plants which run continuously. Nuclear, coal, oil, gas and some hydro plants can supply base load. If well construction costs for natural gas are below $ 10 per MWh, generating electricity from natural gas is cheaper than generating power by burning coal.

In the next part, we are going to talk about the main type of electricity generation by using conventional energy sources, fossil fuels such as coal, oil or gas, also called thermal power.

2. Thermal power generation

Electricity demand varies greatly by season and time of day. Thermal power is a vitally important means of power generation for ensuring a stable supply of electricity by flexibly responding to fluctuations in power demand. Because thermal power generation can flexibly adapt to changes in demand, it plays a central role in maintaining the power supply. There are several types of thermal power generation.

2.1 Steam power generation

Steam power generation (see Fig. 2.2) is a power generation method that takes advantage of the expansive power of steam. Heat produced by burning fuel such as heavy oil, LNG (liquefied natural gas), or coal creates high-temperature, high-pressure steam. This steam is used to turn a steam turbine impeller, causing a generator attached to the turbine to move and generate power. With steam power generation, thermal energy is used in a relatively low-temperature range (600 ℃ or below). The thermal efficiency of steam power generation is in the 41.6% to 45.2% range.

Fuels such as heavy oil, LNG (liquefied natural gas) and coal are burned inside a boiler to generate steam at high temperature and high pressure. This steam is used to rotate the impeller of the steam turbine. This drives the power generators connected to the turbine that generate electricity.

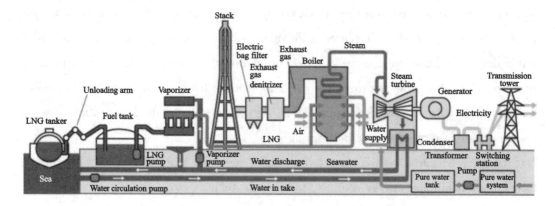

Figure 2. 2 Outline of steam power generation

Source：https：//www. kepco. co. jp/english/corporate/energy/thermal_power/shikumi/index. html.

This system has a thermal efficiency of around 42% to 46% and functions as a base-to-middle-load supply.

2. 2 Combined-cycle power generation

Combined-cycle power generation (see Fig. 2. 3) is a power generation method that combines a gas turbine and steam turbine. It combines gas turbine power generation, by which fuel in compressed air is burned to generate combustion gas whose expansive power is then used to turn a generator; and steam power generation, by which the residual heat from the exhaust gas is recovered to turn a steam turbine. As a result, a high thermal efficiency of 47. 2% can be obtained. In addition, since it consists of a small gas turbine and steam turbine, operation starting and stopping are quick and easy, and fluctuations in demand can be quickly met.

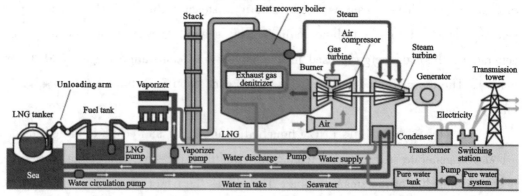

Figure 2. 3 Outline of combined-cycle power generation

Source：https：//www. kepco. co. jp/english/corporate/energy/thermal_power/shikumi/index. html.

This method of generating electricity incorporates a gas turbine whose waste heat is reused to drive a steam turbine. The gas turbine is powered by high-temperature combustion gas that, after being discharged from the gas turbine, is efficiently recovered by means of a heat recovery boiler. This produces steam of sufficient temperature and pressure to drive the steam turbine and generate electricity. This configuration ensures high thermal efficiency, as the cost per unit of power generated is lower than that of oil-fired thermal power. It is used to provide the base-to-middle-load supply.

2.3　Gas turbine power generation

This electricity generating system makes electricity by burning fuels such as LNG (liquefied natural gas) or kerosene to produce high-temperature combustion gases with sufficient energy to rotate a gas turbine (see Fig. 2.4).

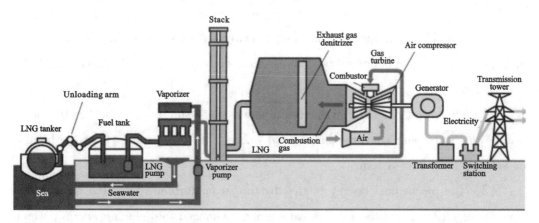

Figure 2.4　Outline of gas turbine power generation

Source: https://www.kepco.co.jp/english/corporate/energy/thermal_power/shikumi/index.html.

3. The generation process and characteristics of thermal power plant

Although there are many types of thermal power plants, from the perspective of energy conversion, the generation process is basically the same, which is the process of converting the chemical energy contained in the fuel (coal) into electric energy (see Fig. 2.5). The whole generation process can be divided into three stages.

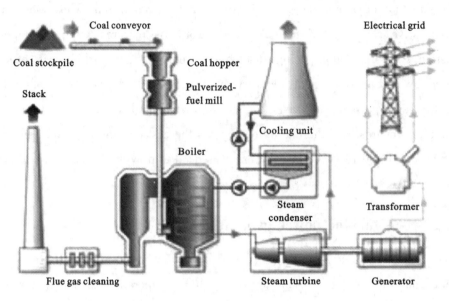

Figure 2.5 How a thermal power plant works

Source: www. quora. com.

(1) The chemical energy of the fuel is converted into heat energy in the boiler, and the water in the boiler is heated to make it into steam, which is called the combustion system.

(2) The steam generated by the boiler enters the steam turbine, pushes the steam turbine to rotate, and converts the thermal energy into mechanical energy, which is called the steam flow system.

(3) The mechanical energy rotated by the steam turbine drives the generator to generate electricity, and the mechanical energy is changed into electric energy, which is called the electrical system.

Compared with hydropower plants and other types of power plants, thermal power plants have the following characteristics.

(1) The layout of the thermal power plant is flexible, and the installed capacity can be determined as required.

(2) The construction period of thermal power plants is short, generally half or even shorter than that of hydropower plants. The one-time construction investment is small, only about half of the hydropower plant.

(3) Coal-fired power plants consume a large amount of coal. At present, the coal used for power generation accounts for about 25% of China's total coal production.

With the transportation cost of coal and large amount of water, the production cost is three to four times higher than that of hydroelectric power generation.

（4）Thermal power plants have various power equipment, complicated control and operation of generator sets, and the power consumption and operating personnel of the plants are more than those of hydropower plants, resulting in high operating costs.

（5）The steam turbine has a long start-up and shutdown process and costs a lot, so it should not be used as a peak-adjusting power supply.

（6）The thermal power plant may cause great pollution to the air and environment.

4. The composition and equipments of thermal power plants

4.1　The composition of thermal power plants

The modern thermal power plant is a large and complex plant that produces electricity and heat. It mainly consists of the following systems: the combustion system, the steam-water system, and the electrical system.

1) Combustion system

The combustion system (see Fig. 2.6) is composed of coal conveying, coal grinding, coarse and fine separation, powder discharging, powder feeding, boiler, dust removal, desulfurization and so on. Coal is conveyed from the coal yard to the coal hopper between the coal silos by the belt conveyor through the electromagnet and coal crusher, and then passes through the coal feeder into the coal mill for grinding. The ground coal powder passes through the hot air from the air preheater and is blown into a coarse and fine separator. The coarse and fine separator sends qualified ground coal power (returns the unqualified pulverized coal to the coal mill), sends it to the powder silo through the powder discharge machine, and feeds the powder coal into the spray. The burner is sent to the boiler for combustion. The flue gas is removed by electric dedusting to remove dust, and then the flue gas is sent to the desulfurization device. The desulfurized gas is sprayed to the flue through the suction fan and discharged into the sky.

27

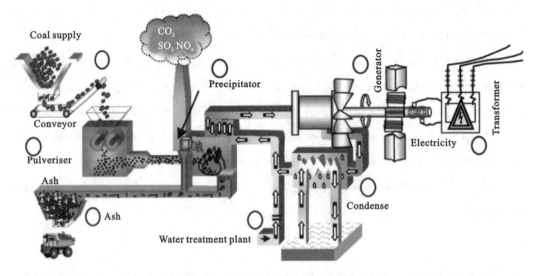

Figure 2. 6   Schematic representation of a coal-fired power plant

Source: https://www. researchgate. net/publication/47517716_Environmental_impacts_of_coal_mining_and_coal_utilization_in_the_UK/figures? lo=1.

2) Steam-water system

The steam-water system (see Fig. 2. 7) of thermal power plant is composed of boiler, steam turbine, condensers, high and low pressure heater, condensate pumps and feedwater pumps. It includes steam-water circulation, chemical water treatment and cooling systems. Water is heated into steam in the boiler, further heated by the heater to become superheated steam, and then enters the turbine through the main steam pipe. As the steam expands, the high-speed flowing steam pushes the turbine's blades to rotate, thereby driving the generator. In order to further improve its thermal efficiency, some of the steam that has been used for work is extracted from some intermediate stages of the steam turbine to heat the feed water. This kind of feedwater regenerative cycle is used in modern large steam turbine sets.

3) Electrical system

Electrical system is mainly composed of secondary exciter, exciter board, main exciter (standby exciter), generator, transformer, high voltage circuit breaker, booster station, distribution device, and etc. The power is generated by the high frequency current generated by the secondary exciter (permanent magnet machine). The current generated by the secondary exciter is rectified by the exciter plate, and then sent to the main exciter. After the primary exciter generates electricity, it is sent to the rotor of the

Figure 2.7　Steam-heated hot water heating system

Source: https://auragmbh.com/industry/steam-heated-hot-water-heating-system/.

generator through the voltage regulator and the deexcitation switch through the carbon brush. The generator rotor rotates its stator coil to induce the current. The strong current passes through the generator outlet in two ways, one way to the factory transformer, the other way to the high-voltage circuit breaker, and the other way to the power grid.

4.2　The main equipments of thermal power plants

In these systems, the main equipments are the boiler, the steam turbine and the generator, which are installed in the main building of the power plant. Main transformers and distribution units are generally installed in independent buildings or outdoors. Among other auxiliary equipments such as water supply system, water supply equipment, water treatment equipment, dust removal equipment, fuel storage and transportation equipment, some are installed in the main building, some are installed in auxiliary buildings or in the open field. By the 1980s, the world's best coal-fired power plants were 40 percent efficient, converting 40 percent of the heat in the fuel into electricity.

1) Boiler

Boiler is one of the main equipments in thermal power plant. Its role is to make fuel combustion in the furnace to generate heat, and heat transfer to the working medium, in order to generate a certain pressure and temperature of steam, for the turbine generator set power generation. Compared with boilers used in other industries, power plant boilers have the characteristics of large capacity, high parameters, complex structure and high degree of automation.

(1) Capacity and parameters of boilers

Boiler capacity is the steam capacity of the boiler, refers to the steam produced by the boiler per hour. The steam capacity that a boiler can achieve is called the rated steam capacity while maintaining the rated steam pressure, rated steam temperature, using the design fuel and achieving the specified thermal efficiency.

The rated parameters of power plant boilers refer to the rated steam pressure and rated steam temperature. The so-called steam pressure and temperature refers to the superheated steam pressure and temperature at the superheater main steam valve outlet. For a boiler with a reheater, the boiler steam parameter should also include the reheat steam parameter.

(2) Classification of power plant boilers

① Classification based on steam parameters

a. Medium pressure boiler. The pressure is 3.822 MPa (39 kgf/cm^2) and the temperature is 450 ℃.

b. High pressure boiler. The pressure is 6-10 MPa, the common pressure is 9.8 MPa (100 kgf/cm^2), and the temperature is 540 ℃.

c. Ultra-high pressure boiler. The pressure is 10-14 MPa, the common pressure is 13.72 MPa (140 kgf/cm^2), the temperature is 555 ℃ or 540 ℃.

d. Subcritical pressure boiler. The pressure is 14-22.2 MPa, the common pressure is 16.66 MPa (170 kgf/cm^2), and the temperature is 555 ℃.

e. Supercritical pressure boiler. The pressure is greater than 22.2 MPa (225.65 kgf/cm^2) and the temperature is 550-570 ℃.

② Classification based on capacity

a. Steam capacity of small boiler is less than 220 t/h.

b. The steam capacity of medium-sized boilers is 220-410 t/h.

c. The evaporation capacity of large boilers shall not be less than 670 t/h.

③ Classification based on burning mode

a. Suspension combustion boiler. The fuel is suspended in the furnace space and can be pulverized coal, oil or gas.

b. Boiling combustion boiler. Solid fuel particles are burned in a boiling state on the grate, also known as a fluidized bed boiler.

④ Classification based on slag discharge method

a. Solid discharge boiler. Ash from fuel combustion is discharged in a solid state.

b. Liquid slag boiler. Ash from fuel combustion is discharged as a liquid.

⑤ Classification based on circulation mode

The boiler can be divided into the following types according to the flow of working fluid in the heating surface of the boiler evaporation.

a. A natural circulation boiler is a boiler with a circulation loop consisting of an oil drum, a descending tube and an ascending tube. It relies on the dead weight difference between the working medium in the descending tube and the ascending tube to produce the power of natural circulation.

b. The forced circulation boiler is equipped with a forced circulation pump on the descending pipe of the circulation circuit to improve the circulation power.

c. Controlling circulation boiler is to install a throttling coil at the entrance of the ascending tube of the forced circulation boiler to control the working fluid flow in each ascending tube to prevent failures such as cycle stagnation or reverse flow.

d. Concurrent boilers are boilers without a circulating loop, and the working fluid passes through each heating surface and becomes superheated steam at one time.

e. Combined circulation boiler, which has a circulation loop and recirculation pump, as well as a switching valve, operates in the recirculation mode at low load and in the DC mode at high load. It can also be cycled at a lower cycle rate under full load.

2) Steam turbine

A steam turbine is a device that extracts thermal energy from pressurized steam and uses it to do mechanical work on a rotating output shaft (see Fig. 2.8). Its modern manifestation was invented by Charles Parsons in 1884.

The steam turbine is a form of heat engine that derives much of its improvement in thermodynamic efficiency from the use of multiple stages in the expansion of the steam, which results in a closer approach to the ideal reversible expansion process. Because the turbine generates rotary motion, it is particularly suited to be used to drive an electrical generator — about 85% of all electricity generation in the United States in the year 2014 was by use of steam turbines.

Figure 2. 8   The rotor of a modern steam turbine used in a power plant

Source: steam turbine-wikipediaen. wikipedia. org.

(1) The working principle

When the steam from the boiler passes through the steam turbine, energy conversion is carried out in the nozzles (stationary blades) and moving blades respectively. Steam turbines can be divided into impulse type and reaction type (see Fig. 2. 9) according to different working principles of steam in moving and static blades.

The working principle of impulse turbine is as follows. Steam with a certain pressure and temperature first expands and accelerates in a stationary nozzle, reducing the pressure and temperature of the steam and converting part of the heat energy into kinetic energy. The high-speed steam flow from the nozzle enters the rotor blade passage installed on the impeller in a certain direction. In the rotor blade passage, the velocity is changed to produce a force, which pushes the impeller and the shaft to rotate, turning the kinetic energy of the steam into the mechanical energy of the shaft.

In a reaction turbine, when the steam flows through the nozzle and moving blade, the steam not only expands and accelerates in the nozzle, but also continues to expand in the moving blade, increasing the velocity of the steam in the moving blade passage. When the steam spurts out from the passage outlet of the moving blade, it gives the moving blade a reverse force. The rotor blades are simultaneously subjected to the impulse force of the steam flow at the nozzle outlet and the counter-force of the steam

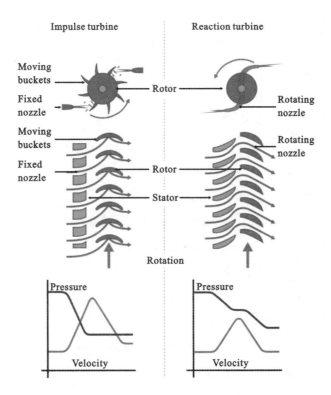

Figure 2. 9   Schematic diagram of an impulse and a reaction turbine

Source: turbines impulse v reaction. svg-wikimedia commons. wikimedia. org.

flow at the nozzle outlet. Under the action of these two forces, the rotor blades drive the impeller and shaft to rotate at high speed, which is the working principle of the reaction turbine.

(2) The composition of steam turbine equipment

Steam turbine equipments include steam turbine body, speed regulation protection and oil system, auxiliary equipments and thermal system.

① Turbine body

The turbine body consists of two parts: stationary and rotating. The former is also known as "stator", including cylinder, separator, nozzle, seal and bearing components; the latter is also known as "rotor", including shaft, impeller and moving blade and other components.

② Speed control protection and oil system

The speed-regulating protection and oil system of steam turbine include governor, oil pump, speed-regulating transmission mechanism, speed-regulating valve, safety protection device and oil cooler.

③ Auxiliary equipments

Auxiliary equipments of steam turbine include condenser, extractor, deaerator, heater and condensate pump.

④ Thermal system

The thermal system of steam turbine includes main steam system, feed water deaeration system, extraction heat recovery system and condensing system.

(3) Classification of steam turbines

① Classification based on working principle

As mentioned above, steam turbines can be divided into impulse type and reaction type according to different working principles.

② Classification based on thermal process characteristics

According to different thermal process characteristics, steam turbines can be divided into the following four types.

a. Condensing turbine. It is characterized by the exhaust steam after work is done in the steam turbine, which enters the condenser and condenses into water under the vacuum state below the atmospheric pressure.

b. Back pressure turbine. Its characteristic is under the circumstance that the exhaust steam pressure is higher than the atmospheric pressure, the exhaust steam will be supplied to the hot user.

c. Intermediate reheat turbine. Its characteristic is in the steam turbine after the high pressure part of the steam extraction, sent to the boiler reheater heating, and then back to the middle pressure part of the steam turbine to continue the work.

d. Extraction turbine. It is characterized by the use of steam heating users with a certain pressure extracted from a certain stage of the turbine, and the exhaust steam still enters the condenser.

③ Classification based on main steam parameters

The steam parameters entering the turbine are steam pressure and temperature. According to different pressure levels, steam turbines can be divided into those kinds.

a. Low pressure steam turbine. The main steam pressure is less than 1.47 MPa.

b. Medium pressure turbine. The main steam pressure is 1.96-3.92 MPa.

c. High pressure steam turbine. The main steam pressure is 5.88-9.8 MPa.

d. Ultra-high pressure steam turbine. The main steam pressure is 11. 77-13. 93 MPa.

e. Subcritical pressure turbine. The main steam pressure is 15. 69-17. 65 MPa.

f. Supercritical pressure turbine. The main steam pressure is greater than 22. 15 MPa.

3）Generator

A generator is a mechanical device that converts other forms of energy into electrical energy. It is driven by a turbine, diesel engine or other power machinery. The energy generated by water flow, air flow, fuel combustion or nuclear fission is converted into mechanical energy and passed to the generator, which is then converted into electrical energy.

A generator is one of the main equipments of the power plant, which is called the three main engines of the thermal power plant together with the boiler and the steam turbine. Generators are widely used in industrial and agricultural production, national defense, science and technology and daily life. There are many forms of generator, but the principle of operation is based on the Faraday law of electromagnetic induction and electromagnetic force. The Faraday law states that whenever a conductor is placed in a varying magnetic field, EMF (electromagnetic field) is induced and this induced EMF is equal to the rate of change of flux linkages. This EMF can be generated when there is either relative space or relative time variation between the conductor and magnetic field. Therefore, the general principle of its construction is that electromagnetic power will be produced with appropriate magnetic and conductive materials to form mutual electromagnetic induction magnetic circuit and circuit, to achieve the purpose of energy conversion.

（1）Working of generators

Generators are basically coils of electric conductors, normally copper wire, that are tightly bound onto a metal core and are mounted to turn around inside an exhibit of large magnets (see Fig. 2. 10-2. 11). An electric conductor moves through a magnetic field, the magnetism will interface with the electrons in the conductor to induce a flow of electrical current inside it.

The conductor coil and its core are called the armature, connecting the armature to the shaft of a mechanical power source, for example an motor, the copper conductor can turn at exceptionally increased speed over the magnetic field.

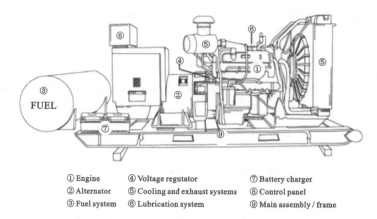

① Engine	④ Voltage regutator	⑦ Battery charger
② Alternator	⑤ Cooling and exhaust systems	⑧ Control panel
③ Fuel system	⑥ Lubrication system	⑨ Main assembly / frame

Figure 2. 10　Design of typical generator

Source：https：//www. researchgate. net/publication/337762805_Combined_heat_and_power/figures？ lo = 1&utm_

source = google&utm_medium = organic.

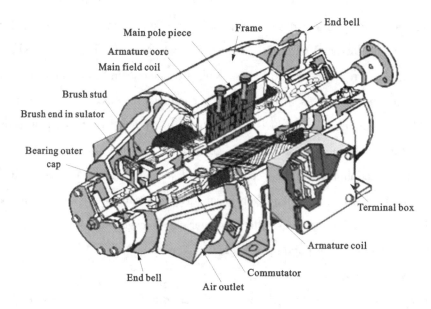

Figure 2. 11　Construction of a generator

Source：http：//www. tpub. com/neets/book5/15i. htm.

When the armature of the generator starts to turn, there will be a weak magnetic field in the iron pole boot. As the armature turns, it starts to raise voltage. Some of this voltage is generated on the excitation windings by the generator regulator. This impressed voltage builds up stronger winding current, raises the strength of the magnetic field. The expanded field produces more voltage in the armature. This, in turn, make more current in the field windings, with a resultant higher armature voltage.

(2) The main structure

The generator usually consists of stator, rotor (see Fig. 2. 12), end cap and bearing.

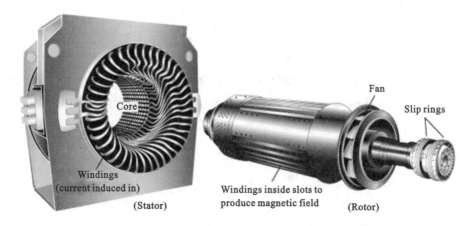

Figure 2. 12 Rotor and stator of electric motor

Source: https://www. researchgate. net/publication/265239692_US_Adoption_of_High-Efficiency_Motors_and_Drives _Lessons_Learned/figures? lo=1&utm_source=google&utm_medium=organic.

The stator is composed of the stator core, the coil winding, the frame and other structural parts fixing these parts.

The rotor consists of rotor core (or magnetic pole, magnetic choke) winding, guard ring, center ring, slip ring, fan and shaft components.

The stator and rotor of the generator are connected and assembled by the bearing and end cover, so that the rotor can rotate in the stator and perform the motion of cutting the magnetic field line, thus generating the induced electric potential.

(3) The classification of generators

Generators are divided into AC and DC generators. There are many types of generators. In principle, it is divided into synchronous generator, asynchronous generator, single-phase generator and three-phase generator. They can also be divided into turbo-generator, hydro-generator, diesel generator, gasoline generator and so on.

① AC generators

AC generators are also called as alternators. It is the most important means of producing electrical power in many of the places since nowadays all the consumers are using AC. It works on the principle of electromagnetic induction. AC generators have two types, one is induction generator and the other is synchronous generator. The induction generator requires no separate DC excitation, regulator controls, frequency control or governor. This concept takes place when conductor coils turn in a magnetic field actuating a current and a voltage. The generators should run at a consistent speed to convey a stable AC voltage, even no load is accessible.

Synchronous generators are large size generators mainly used in power plants. These may be rotating field type or rotating armature type. In rotating armature type, armature is at rotor and field is at stator. Rotor armature current is taken through slip rings and brushes. These are limited due to high wind losses. These are used for low power output applications. Rotating field type of alternator is widely used because of high power generation capability and absence of slip rings and brushes.

Generators can be either three-phase or two-phase generators. A two-phase alternator produces two completely separate voltages. Each voltage may be considered as a single-phase voltage. Each generates a voltage that is completely independent of the other. The three-phase alternator has three single-phase windings spaced such that the voltage induced in any one phase is displaced at an angle of 120 degrees from the other two. These can be connected either in Delta or Wye Connections. In Delta Connection each coil end is connected together to form a closed loop. A Delta Connection appears like the Greek letter Delta (Δ). In Wye Connection one end of each coil connected together and the other end of each coil left open for external connections. A Wye Connection appears as the letter Y.

These generators are packaged with an engine or turbine to be used as a motor-generator set and used in applications like naval, oil and gas extraction, mining machinery, wind power plants, and etc.

AC generators have the following advantages.

a. These generators are generally maintenance free, because of absence of brushes.

b. These generators are easily stepping up and stepping down through transformers.

c. Transmission link size might be thinner because of stepping up feature.

d. The size of the generator is relatively smaller than that of DC generator.

e. Losses are relatively less than DC generator.

f. These generator breakers are relatively smaller than DC breakers.

② DC generators

DC generator is typically found in off-grid applications. These generators give a seamless power supply directly into electric storage devices and DC power grids without novel equipment. The stored power is carried to loads through DC-AC converters. The DC generators could be controlled back to an unmoving speed as batteries tend to be stimulating to recover considerably more fuel.

DC generators are classified according to the way their magnetic field is developed in the stator of the machine into permanent-magnet DC generators, separately-excite DC generators and self-excited DC generators.

Permanent magnet DC generators do not require external field excitation because it has permanent magnets to produce the flux. These are used for low power applications like dynamos. Separately-excite DC generators require external field excitation to produce the magnetic flux. We can also vary the excitation to get variable output power. These are used in electro plating and electro refining applications. Due to residual magnetism present in the poles of the stator self-excited DC generators can able to produce their own magnetic field once it is started. These are simple in design and no need to have the external circuit to vary the field excitation. Again these self-excited DC generators are classified into shunt, series, and compound generators.

Advantages of DC generators are as followed.

a. Mainly DC machines have the wide variety of operating characteristics which can be obtained by selection of the method of excitation of the field windings.

b. The output voltage can be smoothed by regularly arranging the coils around the armature. This leads to less fluctuations which is desirable for some steady state applications.

c. No shielding need for radiation, so cable cost will be less as compared to AC.

 New words and phrases

AC　*abbr.*　交流电（alternating current）

ampere   *n.*   安培（计算电流强度的标准单位）

armature   *n.*   电枢

bearing   *n.*   轴承

boiler   *n.*   锅炉

booster station   升压站；增压站

coal feeder   给煤机；供煤机

coal hopper   煤斗；煤炭漏斗

coarse   *adj.*   粗糙的

concurrent boiler   直流锅炉

condenser   *n.*   冷凝器

condensing turbine   冷凝式汽轮机

configuration   *n.*   配置；结构

correlate   *v.*   关联

coulomb   *n.*   库仑（电量单位）

current   *n.*   （水，气，电）流；趋势

DC   *abbr.*   直流电（direct current）

dead weight   自重；净重

dedust   *v.*   除尘；去灰

desulfurization   *n.*   脱硫作用

electric motor   电动机

electric potential   电位；电势

electrical system   电气系统

electromechanical   *adj.*   电动机械的；机电的

electron   *n.*   电子

EMF   *abbr.*   电磁场（electromagnetic field）

exciter   *n.*   励磁机

exhaust gas   排出的气体；废气

extraction turbine   抽汽式汽轮机

feedwater   *n.*   给水；锅炉给水

fluctuation   *n.*   起伏；波动

fluidized bed   流化床

grind   *v.*   磨碎

impulse turbine   冲动式汽轮机

induce   *v.*   诱导;引起

installed capacity   装机容量;装置容量

kerosene   *n.*   煤油;火油

nozzle   *n.*   喷嘴;管口

propel   *v.*   推进;驱使

pulverized   *adj.*   粉状的;成粉末的

rated   *adj.*   额定的

reaction turbine   反动式汽轮机

rectify   *v.*   整顿;纠正

regenerative   *adj.*   再生的;更生的

residual   *adj.*   剩余的;残留的

rotor   *n.*   转子

slag discharge   排渣

stator   *n.*   定子

steam turbine   汽轮机;蒸汽轮机

subcritical   *adj.*   亚临界的;次临界的

synchronous   *adj.*   同步的;同时的

thermal efficiency   热效率

thermal power plant   火电厂;火力发电厂

thermodynamic efficiency   热力学效率

voltage regulator   稳压器;调压器

working medium   工作介质;使用介质

 Exercises

I . Warming-up questions

Directions：Give brief answers to the following questions.

1. What do you know about electric power?

2. How is electricity generated?

3. What are the main components of thermal power plants?

Ⅱ. Technical terms

Directions：Please give the Chinese or English equivalents of the following terms.

1. desulfurization
2. feedwater
3. impulse turbine
4. electric potential
5. exhaust gas
6. alternating current
7. 火电厂
8. 工作介质
9. 直流电
10. 排渣
11. 悬浮燃烧锅炉
12. 冷凝式汽轮机

Ⅲ. Blank filling

Directions：Complete the following sentences by choosing words or phrases given below.

fluctuation	induce	dedust	thermal efficiency	regenerative

1. The separation efficiency of the original _____ system for the raw gas is bad，so that many difficulties are come to the normal operation of the CO shift system.

2. To recreate this unique _____ ability，the MIT team devised a novel set of self-assembling molecules that use photons to shake electrons loose in the form of electricity.

3. Japanese researchers discovered that hot alcoholic beverages _____ superconductivity in iron-based compounds.

4. At the moment almost 80 percent of the turbines and platforms have to be imported which means they are susceptible to currency _____.

5. The improvement in _____ for a simple Rankine cycle is by virtue of the bled-steam releasing all of its heat to the feed-water，and little or none to the condenser.

Ⅳ. Translation

Directions：Translate the following passages from English into Chinese or from Chinese into English.

Passage 1

In an organic Rankine cycle (ORC) system, the generator is directly coupled to the turbine expander that is designed for a high-speed drive so as to reduce the size and increase the efficiency. Additionally, the generator rotor will operate at variable speeds according to the operating conditions of the ORC system. However, due to the ripple or fluctuation of the turbine rotation speed caused by the unpredictable nature of the ORC system, the generator is exposed to the speed ripple, which in turn causes significant vibration and noise.

Passage 2

锅炉是一种能量转换设备,它是利用燃料燃烧释放的热能或其他热能将工质水或其他流体加热到一定参数的设备。锅原义是指在火上加热的盛水容器,炉是指燃烧燃料的场所,锅炉包括锅和炉两大部分。锅炉中产生的热水或蒸汽可直接为工业生产和人民生活提供所需热能,也可通过蒸汽动力装置转换为机械能,或再通过发电机将机械能转换为电能。提供热水的锅炉称为热水锅炉,主要用于日常生活中,工业生产中也有少量应用。产生蒸汽的锅炉称为蒸汽锅炉,常简称为锅炉,多用于火电站、船舶、机车和工矿企业。

Additional reading

Unit 3
Wind Power Generation

1. Understanding wind power generation

Wind power or wind energy is the use of wind to provide the mechanical power through wind turbines to turn electric generators and traditionally to do other work, like milling or pumping. Wind power is a sustainable and renewable energy, and has a much smaller impact on the environment compared to burning fossil fuels.

Wind farms consist of many individual wind turbines, which are connected to the electric power transmission network. Onshore wind is an inexpensive source of electric power, competitive with or in many places cheaper than coal or gas plants. Onshore wind farms also have an impact on the landscape, as typically they need to be spread over more land than other power stations and need to be built in wild and rural areas, which can lead to "industrialization of the countryside" and habitat loss. Offshore wind is steadier and stronger than on land and offshore farms have less visual impact, but construction and maintenance costs are higher. Small onshore wind farms can feed some energy into the grid or provide electric power to isolated off-grid locations.

Wind is an intermittent energy source, which cannot make electricity nor be dispatched on demand. It also gives variable power, which is consistent from year to year but varies greatly over shorter time scales. Therefore, it must be used together with other electric power sources or storage to give a reliable supply. As the proportion of wind power in a region increases, more conventional power sources are needed to back it up (such as fossil fuel power and nuclear power), and the grid may need to be upgraded. Power-management techniques such as having dispatchable power sources, enough hydroelectric power, excess capacity, geographically distributed turbines, exporting and importing power to neighboring areas, energy storage, or reducing demand when wind production is low, can in many cases overcome these problems. Weather forecasting permits the electric-power network to be readied for the predictable variations in production in case that occur.

In 2018, global wind power capacity grew 9. 6% to 591 GW and yearly wind energy production grew 10%, reaching 4. 8% of worldwide electric power usage, and providing 14% of the electricity in the European Union.

Denmark is the country with the highest penetration of wind power, with 43. 4% of its consumed electricity from wind in 2017. At least 83 other countries are using wind power to supply their electric power grids.

2. History of wind power generation

Wind power has been used as long as humans have put sails into the wind. Wind-powered machines used to grind grain and pump water, the windmill and wind pump, were developed in what is now Iran, Afghanistan and Pakistan by the 9th century. Wind power was widely available and not confined to the banks of fast-flowing streams, or later, requiring sources of fuel. Wind-powered pumps drained the polders of the Netherlands, and in arid regions such as the American mid-west or the Australian outback, wind pumps provided water for livestock and steam engines.

The first windmill used for the production of electric power was built in Scotland in July 1887 by Prof James Blyth of Anderson's College, Glasgow (the precursor of Strathclyde University). Blyth's 10 metres high, cloth-sailed wind turbine was installed in the garden of his holiday cottage at Marykirk in Kincardineshire and was used to charge accumulators developed by the Frenchman Camille Alphonse Faure, to power the lighting in the cottage, thus making it the first house in the world to have its electric power supplied by wind power. Blyth offered the surplus electric power to the people of Marykirk for lighting the main street, however, they turned down the offer as they thought electric power was "the work of the devil". Although he later built a wind turbine to supply emergency power to the local Lunatic Asylum, Infirmary and Dispensary of Montrose, the invention never really caught on as the technology was not considered to be economically viable.

Across the Atlantic, in Cleveland, Ohio, a larger and heavily engineered machine was designed and constructed in the winter of 1887-1888 by Charles F. Brush. This was built by his engineering company at his home and operated from 1886 until 1900. The Brush wind turbine had a rotor 17 metres in diameter and was mounted on an 18 metres tower. Although large by today's standards, the machine was only rated at 12 kW. The connected dynamo was used either to charge a bank of batteries or to operate up to 100 incandescent light bulbs, three arc lamps, and various motors in Brush's laboratory.

With the development of electric power, wind power found new applications in lighting buildings remote from centrally-generated power. Throughout the 20th century parallel paths developed small wind stations suitable for farms or residences. The 1973 oil crisis triggered investigation in Denmark and the United States that led to larger utility-scale wind generators that could be connected to electric power grids for remote use of power. By 2008, the US installed capacity had reached 25.4 gigawatts, and by 2012 the installed capacity was 60 gigawatts. Today, wind powered generators operate in every size range between tiny stations for battery charging at isolated residences, up to near-gigawatt sized offshore wind farms that provide electric power to national electrical networks.

3. Generation and transmission of electric power

3.1 Turbine design

Wind turbines are devices that convert the wind's kinetic energy into electrical power. The result of over a millennium of windmill development and modern engineering, today's wind turbines are manufactured in a wide range of horizontal axis and vertical axis types. The smallest turbines are used for applications such as battery charging for auxiliary power. Slightly larger turbines can be used for making small contributions to a domestic power supply while selling unused power back to the utility supplier via the electrical grid. Arrays of large turbines, known as wind farms, have become an increasingly important source of renewable energy and are used in many countries as part of a strategy to reduce their reliance on fossil fuels. The figure shows the typical wind turbine components (see Fig. 3.1).

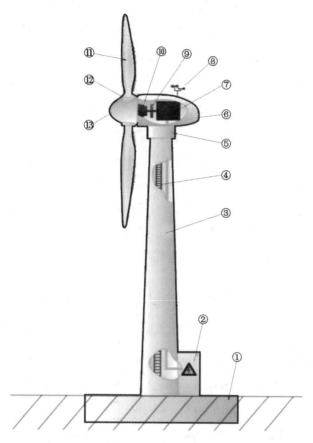

① Foundation　② Connection to the electric grid　③ Tower　④ Access ladder　⑤ Wind orientation control (Yaw control)
⑥ Nacelle　⑦ Generator　⑧ Anemometer　⑨ Electric or mechanical brake　⑩ Gearbox
⑪ Rotor blade　⑫ Blade pitch control　⑬ Rotor hub

Figure 3.1　Typical wind turbine components

Source: https://en.wikipedia.org/wiki/Wind_power#References.

Wind turbine design is the process of defining the form and specifications of a wind turbine to extract energy from the wind. A wind turbine installation consists of the necessary systems needed to capture the wind's energy, point the turbine into the wind, convert mechanical rotation into electrical power, and other systems to start, stop, and control the turbine.

In 1919, the German physicist Albert Betz showed that for a hypothetical ideal wind-energy extraction machine, the fundamental laws of conservation of mass and energy allowed no more than 16/27 (59%) of the kinetic energy of the wind to be captured. This Betz limit can be approached in modern turbine designs, which may reach 70% to 80% of the theoretical Betz limit.

The aerodynamics of a wind turbine is not straightforward. The air flow at the blades is not the same as the airflow far away from the turbine. The very nature of the way in which energy is extracted from the air also causes air to be deflected by the turbine. In addition, the aerodynamics of a wind turbine at the rotor surface exhibit phenomena that are rarely seen in other aerodynamic fields. The shape and dimensions of the blades of the wind turbine are determined by the aerodynamic performance required to efficiently extract energy from the wind, and by the strength required to resist the forces on the blade.

In addition to the aerodynamic design of the blades, the design of a complete wind power system must also address the design of the installation's rotor hub, nacelle, tower structure, generator, controls, and foundation.

3.2 Onshore and offshore wind farms

A wind farm is a group of wind turbines in the same location used for production of electric power. A large wind farm may consist of several hundred individual wind turbines distributed over an extended area, but the land between the turbines may be used for agricultural or other purposes (see Fig. 3.2). For example, Gansu Wind Farm, the largest wind farm in the world, has several thousand turbines. A wind farm may also be located offshore.

Almost all large wind turbines have the same design — a horizontal axis wind turbine having an upwind rotor with three blades, attached to a nacelle on top of a tall tubular tower.

In a wind farm, individual turbines are interconnected with a medium voltage (often 34.5 kV) power collection system and communications network. In general, a distance of 7D (seven times the rotor diameter of the wind turbine) is set between each turbine in a fully developed wind farm. At a substation, this medium-voltage electric current is increased in voltage with a transformer for connection to the high voltage electric power transmission system.

Figure 3. 2   Wind power stations

Source: https://pixabay. com/zh/photos/wind-farm-power-station-wind-power-2856793/.

Induction generators, which were often used for wind power projects in the 1980s and 1990s, require reactive power for excitation, so substations used in wind-power collection systems include substantial capacitor banks for power factor correction. Different types of wind turbine generators behave differently during transmission grid disturbances, so extensive modelling of the dynamic electromechanical characteristics of a new wind farm is required by transmission system operators to ensure predictable stable behaviour during system faults. In particular, induction generators cannot support the system voltage during faults, unlike steam or hydro turbine-driven synchronous generators.

Induction generators aren't used in current turbines. Instead, most turbines use variable speed generators combined with partial- or full-scale power converter between the turbine generator and the collector system, which generally have more desirable properties for grid interconnection and have low voltage ride through-capabilities. Modern concepts use either doubly fed machines with partial-scale converters or squirrel-cage induction generators or synchronous generators (both permanently and electrically excited) with full scale converters.

Offshore wind power refers to the construction of wind farms in large bodies of water to generate electric power (see Fig. 3. 3). These installations can utilize the more frequent and powerful winds that are available in these locations and have less

aesthetic impact on the landscape than land based projects. However, the construction and the maintenance costs are higher.

Figure 3. 3　Offshore wind farm in the North Sea

Source: https://en. wikipedia. org/wiki/Wind_power#References.

In 2012, 1,662 turbines at 55 offshore wind farms in 10 European countries produced 18 TWh, enough to power almost five million households. As of September 2018 the Walney Extension in the United Kingdom is the largest offshore wind farm in the world at 659 MW.

3.3　Collection and transmission of electricity

In a wind farm, individual turbines are interconnected with a medium voltage (usually 34.5 kV) power collection system and communications network. At a substation, this medium-voltage electric current is increased in voltage with a transformer for connection to the high voltage electric power transmission system.

A transmission line is required to bring the generated power to (often remote) markets. For an off-shore station this may require a submarine cable. Construction of a new high-voltage line may be too costly for the wind resource alone, but wind sites may take advantage of lines installed for conventionally fueled generation.

4. Some parameters in wind power generation

4.1   Capacity factor

Since wind speed is not constant, a wind farm's annual energy production is never as much as the sum of the generator nameplate ratings multiplied by the total hours in a year. The ratio of actual productivity in a year to this theoretical maximum is called the capacity factor. Typical capacity factors are 15%-50%; values at the upper end of the range are achieved in favorable sites and are due to wind turbine design improvements.

Online data is available for some locations, and the capacity factor can be calculated from the yearly output. For example, the German nationwide average wind power capacity factor over all of 2012 was just under 17.5% (45,867 GWh/yr/ (29.9 GW × 24 × 366) = 0.1746), and the capacity factor for Scottish wind farms averaged 24% between 2008 and 2010.

Unlike fueled generating plants, the capacity factor is affected by several parameters, including the variability of the wind at the site and the size of the generator relative to the turbine's swept area. A small generator would be cheaper and achieve a higher capacity factor but would produce less electric power (and thus less profit) in high winds. Conversely, a large generator would cost more but generate little extra power and, depending on the type, may stall out at low wind speed. Thus an optimum capacity factor of around 40%-50% would be aimed for.

A 2008 study released by the US Department of Energy noted that the capacity factor of new wind installations was increasing as the technology improves, and projected further improvements for future capacity factors. In 2010, the department estimated the capacity factor of new wind turbines in 2010 to be 45%. The annual average capacity factor for wind generation in the US has varied between 29.8% and 34% during the period 2010-2015.

4.2   Penetration

Wind energy penetration is the fraction of energy produced by wind compared with the total generation. Wind power's share of worldwide electricity usage at the end of

2019 was 4.8%, up from 3.5% in 2015.

There is no generally accepted maximum level of wind penetration. The limit for a particular grid will depend on the existing generating plants, pricing mechanisms, capacity for energy storage, demand management and other factors. An interconnected electric power grid will already include reserve generating and transmission capacity to allow for equipment failures. This reserve capacity can also serve to compensate for the varying power generation produced by wind stations. Studies have indicated that 20% of the total annual electrical energy consumption may be incorporated with minimal difficulty. These studies have been for locations with geographically dispersed wind farms, some degree of dispatchable energy or hydropower with storage capacity, demand management, and interconnected to a large grid area enabling the export of electric power when needed. Beyond the 20% level, there are few technical limits, but the economic implications become more significant. Electrical utilities continue to study the effects of large scale penetration of wind generation on system stability and economics.

A wind energy penetration figure can be specified for different duration of time, but is often quoted annually. To obtain 100% from wind annually requires substantial long term storage or substantial interconnection to other systems which may already have substantial storage. On a monthly, weekly, daily, or hourly basis — or less — wind might supply as much as or more than 100% of current use, with the rest stored or exported. Seasonal industry might then take advantage of high wind and low usage times such as at night when wind output can exceed normal demand. Such industry might include production of silicon, aluminum, steel, or of natural gas, and hydrogen, and using future long term storage to facilitate 100% energy from variable renewable energy. Homes can also be programmed to accept extra electric power on demand, for example, by remotely turning up water heater thermostats.

In Australia, the state of South Australia generates around half of the nation's wind power capacity. By the end of 2011 wind power in South Australia, championed by Premier (and Climate Change Minister) Mike Rann, reached 26% of the State's electric power generation, edging out coal for the first time. At this stage South Australia, with only 7.2% of Australia's population, had 54% of Australia's installed capacity.

4.3   Variability

Electric power generated from wind power can be highly variable at several different timescales: hourly, daily, or seasonally. Annual variation also exists, but is not as significant. Because instantaneous electrical generation and consumption must remain in balance to maintain grid stability, this variability can present substantial challenges to incorporating large amounts of wind power into a grid system. Intermittency and the non-dispatchable nature of wind energy production can raise costs for regulation, incremental operating reserve, and (at high penetration levels) could require an increase in the already existing energy demand management, load shedding, storage solutions or system interconnection with HVDC cables.

Fluctuations in load and allowance for failure of large fossil-fuel generating units require operating reserve capacity, which can be increased to compensate for variability of wind generation.

Wind power is variable, and during low wind periods it must be replaced by other power sources. Transmission networks presently cope with outages of other generation plants and daily changes in electrical demand, but the variability of intermittent power sources such as wind power, is more frequent than those of conventional power generation plants which, when scheduled to be operating, may be able to deliver their nameplate capacity around 95% of the time.

Presently, grid systems with large wind penetration require a small increase in the frequency of usage of natural gas spinning reserve power plants to prevent a loss of electric power in the event that there is no wind. At low wind power penetration, this is less of an issue.

GE has installed a prototype wind turbine with onboard battery similar to that of an electric car, equivalent of 1 minute of production. Despite the small capacity, it is enough to guarantee that power output complies with forecast for 15 minutes, as the battery is used to eliminate the difference rather than provide full output. In certain cases the increased predictability can be used to take wind power penetration from 20 to 30 or 40 percent. The battery cost can be retrieved by selling burst power on demand and reducing backup needs from gas plants.

In the UK, there were 124 separate occasions from 2008 to 2010 when the nation's wind output fell to less than 2% of installed capacity. A report on Denmark's wind power noted that their wind power network provided less than 1% of average demand on 54 days during the year 2002. Wind power advocates argue that these periods of low wind can be dealt with by simply restarting existing power stations that have been held in readiness, or interlinking with HVDC. Electrical grids with slow-responding thermal power plants and without ties to networks with hydroelectric generation may have to limit the use of wind power. According to a 2007 Stanford University study published in the *Journal of Applied Meteorology and Climatology*, interconnecting ten or more wind farms can allow an average of 33% of the total energy produced (i. e. about 8% of total nameplate capacity) to be used as reliable, baseload electric power which can be relied on to handle peak loads, as long as minimum criteria are met for wind speed and turbine height.

Conversely, on particularly windy days, even with penetration levels of 16%, wind power generation can surpass all other electric power sources in a country. In Spain, in the early hours of 16 April 2012 wind power production reached the highest percentage of electric power production till then, at 60. 46% of the total demand. In Denmark, which had power market penetration of 30% in 2013, over 90 hours, wind power generated 100% of the country's power, peaking at 122% of the country's demand at 2 am on 28 October.

A 2006 International Energy Agency forum presented costs for managing intermittency as a function of wind-energy's share of total capacity for several countries. Three reports on the wind variability in the UK issued in 2009, generally agree that variability of wind needs to be taken into account by adding 20% to the operating reserve, but it does not make the grid unmanageable. The additional costs, which are modest, can be quantified.

The combination of diversifying variable renewables by type and location, forecasting their variation, and integrating them with dispatchable renewables, flexible fueled generators, and demand response can create a power system that has the potential to meet power supply needs reliably.

Solar power tends to be complementary to wind. On daily to weekly timescales, high pressure areas tend to bring clear skies and low surface winds, whereas low pressure areas tend to be windier and cloudier. On seasonal timescales, solar energy peaks in summer, whereas in many areas wind energy is lower in summer and higher in winter. Thus the seasonal variation of wind and solar power tend to cancel each other somewhat. In 2007 the Institute for Solar Energy Supply Technology of the University of Kassel pilot-tested a combined power plant linking solar, wind, biogas and hydrostorage to provide load-following power around the clock and throughout the year, entirely from renewable sources.

4.4　Predictability

Wind power forecasting methods are used, but predictability of any particular wind farm is low for short-term operation. For any particular generator there is an 80% chance that wind output will change less than 10% in an hour and a 40% chance that it will change 10% or more in 5 hours.

However, studies by Graham Sinden (2009) suggest that, in practice, the variations in thousands of wind turbines, spread out over several different sites and wind regimes, are smoothed. As the distance between sites increases, the correlation between wind speeds measured at those sites, decreases.

Thus, while the output from a single turbine can vary greatly and rapidly as local wind speeds vary, as more turbines are connected over larger and larger areas the average power output becomes less variable and more predictable.

Wind power hardly ever suffers major technical failures, since failures of individual wind turbines have hardly any effect on overall power, so that the distributed wind power is reliable and predictable, whereas conventional generators, while far less variable, can suffer major unpredictable outages.

4.5　Energy storage

Typically, conventional hydroelectricity complements wind power very well. When the wind is blowing strongly, nearby hydroelectric stations can temporarily hold back their water. When the wind drops they can, provided they have the generation capacity, rapidly increase production to compensate. This gives a very even overall power supply and virtually no loss of energy and uses no more water.

Alternatively, where a suitable head of water is not available, pumped-storage hydroelectricity or other forms of grid energy storage such as compressed air energy

storage and thermal energy storage can store energy developed by high-wind periods and release it when needed. The type of storage needed depends on the wind penetration level — low penetration requires daily storage, and high penetration requires both short and long term storage — as long as a month or more. Stored energy increases the economic value of wind energy since it can be shifted to displace higher cost generation during peak demand periods. The potential revenue from this arbitrage can offset the cost and losses of storage. For example, in the UK, the 1.7 GW Dinorwig pumped-storage plant evens out electrical demand peaks, and allows base-load suppliers to run their plants more efficiently. Although pumped-storage power systems are only about 75% efficient, and have high installation costs, their low running costs and ability to reduce the required electrical base-load can save both fuel and total electrical generation costs.

In particular geographic regions, peak wind speeds may not coincide with peak demand for electrical power. In the US states of California and Texas, for example, hot days in summer may have low wind speed and high electrical demand due to the use of air conditioning. Some utilities subsidize the purchase of geothermal heat pumps by their customers, to reduce electric power demand during the summer months by making air conditioning up to 70% more efficient; widespread adoption of this technology would better match electric power demand to wind availability in areas with hot summers and low summer winds. A possible future option may be to interconnect widely dispersed geographic areas with an HVDC "super grid". In the US it is estimated that to upgrade the transmission system to take in planned or potential renewables would cost at least $60 bn, while the society value of added wind power would be more than that cost.

Germany has an installed capacity of wind and solar that can exceed daily demand, and has been exporting peak power to neighboring countries, with exports which amounted to some 14.7 billion kWh in 2012. A more practical solution is the installation of thirty days storage capacity able to supply 80% of demand, which will become necessary when most of Europe's energy is obtained from wind power and solar power. Just as the EU requires member countries to maintain 90 days of oil it can be strategic reserves expected that countries will provide electric power storage, instead of expecting to use their neighbors for net metering.

4.6   Capacity credit, fuel savings and energy payback

The capacity credit of wind is estimated by determining the capacity of conventional plants displaced by wind power, whilst maintaining the same degree of system security. According to the American Wind Energy Association, production of wind power in the United States in 2015 avoided consumption of 280 million cubic metres of water and reduced CO_2 emissions by 132 million metric tons, while providing $7.3 bn in public health savings.

The energy needed to build a wind farm divided into the total output over its life, energy return on energy invested (EROI), of wind power varies but averages about 20-25. Thus, the energy payback time is typically around a year.

5. Benefits and costs of wind power

Wind power is capital intensive, but has no fuel costs. The price of wind power is therefore much more stable than the volatile prices of fossil fuel sources. The marginal cost of wind energy once a station is constructed is usually less than 1 cent per kWh.

However, the estimated average cost per unit of electric power must incorporate the cost of construction of the turbine and transmission facilities, borrowed funds, return to investors (including cost of risk), estimated annual production, and other components, averaged over the projected useful life of the equipment, which may be in excess of twenty years. Energy cost estimates are highly dependent on these assumptions so published cost figures can differ substantially. In 2004, wind energy cost a fifth of what it did in the 1980s, and some expected that downward trend to continue as larger multi-megawatt turbines were mass-produced. In 2012, capital costs for wind turbines were substantially lower than 2008-2010 but still above 2002 levels. A 2011 report from the American Wind Energy Association stated, "wind's costs have dropped over the past two years, in the range of 5 to 6 cents per kilowatt-hour recently... about 2 cents cheaper than coal-fired electric power, and more projects were financed through debt arrangements than tax equity structures last year... winning more mainstream acceptance from Wall Street's banks.... Equipment makers can also deliver products in the same year that they are ordered instead of waiting up to three years as was the case in previous cycles... 5,600 MW of new installed capacity is under construction in the United States, more than double the number at this point in 2010. Thirty-five percent of all new power generation built in the United States since 2005 has come from wind, more than new gas and coal plants combined, as power providers are increasingly enticed to wind as a convenient hedge against unpredictable commodity price moves."

A British Wind Energy Association report gives an average generation cost of onshore wind power of around 3.2 pence (between US 5 and 6 cents) per kWh (2005). Cost per unit of energy produced was estimated in 2006 to be 5 to 6 percent above the cost of new generating capacity in the US for coal and natural gas: wind cost was estimated at $55.80 per MWh, coal at $53.10/MWh and natural gas at $52.50. Similar comparative results with natural gas were obtained in a governmental study in the UK in 2011. In 2011, power from wind turbines could be already cheaper than fossil or nuclear plants; it is also expected that wind power will be the cheapest form of energy generation in the future. The presence of wind energy, even when subsidised, can reduce costs for consumers (€5 billion/yr in Germany) by reducing the marginal price, by minimising the use of expensive peaking power plants.

A 2012 EU study shows base cost of onshore wind power is similar to coal, when subsidies and externalities are disregarded. Wind power has some of the lowest external costs.

In February 2013, Bloomberg New Energy Finance (BNEF) reported that the cost of generating electric power from new wind farms is cheaper than new coal or new baseload gas plants. When including the current Australian federal government carbon pricing scheme their modeling gives costs (in Australian dollars) of $80/MWh for new wind farms, $143/MWh for new coal plants and $116/MWh for new baseload gas plants. The modeling also shows that "even without a carbon price (the most efficient way to reduce economy-wide emissions) wind energy is 14% cheaper than new coal and 18% cheaper than new gas". Part of the higher costs for new coal plants is due to high financial lending costs because of "the reputational damage of emissions-intensive investments". The expense of gas fired plants is partly due to "export market" effects on local prices. Costs of production from coal fired plants built in "the 1970s and 1980s" are cheaper than renewable energy sources because of depreciation. In 2015

BNEF calculated the levelized cost of electricity (LCOE) per MWh in new power plants (excluding carbon costs): $85 for onshore wind ($175 for offshore), $66-75 for coal in the Americas ($82-105 in Europe), gas $80-100. A 2014 study showed unsubsidized LCOE costs between $37-81, depending on region. A 2014 US DOE report showed that in some cases power purchase agreement prices for wind power had dropped to record lows of $23.5/MWh.

The cost has reduced as wind turbine technology has improved. There are now longer and lighter wind turbine blades, improvements in turbine performance and increased power generation efficiency. Also, wind project capital and maintenance costs have continued to decline. For example, the wind industry in the US in early 2014 were able to produce more power at lower cost by using taller wind turbines with longer blades, capturing the faster winds at higher elevations. This has opened up new opportunities and in Indiana, Michigan, and Ohio, the price of power from wind turbines built 90-120 metres above the ground can since 2014 compete with conventional fossil fuels like coal. Prices have fallen to about 4 cents per kilowatt-hour in some cases and utilities have been increasing the amount of wind energy in their portfolio, saying it is their cheapest option.

A number of initiatives are working to reduce costs of electric power from offshore wind. One example is the Carbon Trust Offshore Wind Accelerator, a joint industry project, involving nine offshore wind developers, which aims to reduce the cost of offshore wind by 10% by 2015. It has been suggested that innovation at scale could deliver 25% cost reduction in offshore wind by 2020. Henrik Stiesdal, former Chief Technical Officer at Siemens Wind Power, has stated that by 2025 energy from offshore wind will be one of the cheapest, scalable solutions in the UK, compared to other renewables and fossil fuel energy sources, if the true cost to society is factored into the cost of energy equation. He calculates the cost at that time to be € 43/MWh for onshore, and € 72/MWh for offshore wind.

In August 2017, the Department of Energy's National Renewable Energy Laboratory (NREL) published a new report on a 50% reduction in wind power cost by 2030. The NREL is expected to achieve advances in wind turbine design, materials and controls to unlock performance improvements and reduce costs. According to international surveyors, this study shows that cost cutting is projected to fluctuate

between 24% and 30% by 2030. In more aggressive cases, experts estimate cost reduction up to 40 percent if the research and development and technology programs result in additional efficiency.

In 2018 a Lazard study found that "the low end levelized cost of onshore wind-generated energy is $29/MWh, compared to an average illustrative marginal cost of $36/MWh for coal", and noted that the average cost had fallen by 7% in a year.

6. Environmental impacts

The environmental impact of wind power is considered to be relatively minor compared to that of fossil fuels. According to the IPCC, in assessments of the life-cycle greenhouse-gas emissions of energy sources, wind turbines have a median value of 12 and 11 (gCO_2eq/kWh) for offshore and onshore turbines, respectively. Compared with other low carbon power sources, wind turbines have some of the lowest global warming potential per unit of electrical energy generated.

Onshore wind farms can have a significant impact on the landscape. Their network of turbines, access roads, transmission lines and substations can result in "energy sprawl". Wind farms typically need to cover more land and be more spread out than other power stations. To power major cities by wind alone would require building wind farms bigger than the cities themselves. A report by the Mountaineering Council of Scotland concluded that wind farms have a negative impact on tourism in areas known for natural landscapes and panoramic views. However, land between the turbines and roads can still be used for agriculture.

Wind farms are typically built in wild and rural areas, which can lead to "industrialization of the countryside" and habitat loss. Habitat loss and habitat fragmentation are the greatest impact of wind farms on wildlife. There are also reports of higher bird and bat mortality at wind turbines as there are around other artificial structures. The scale of the ecological impact may or may not be significant, depending on specific circumstances. Prevention and mitigation of wildlife fatalities, and protection of peat bogs, affect the siting and operation of wind turbines.

Wind turbines generate noise. At a residential distance of 300 metres this may be around 45 dB, which is slightly louder than a refrigerator. At 1.5 km distance they become inaudible. There are anecdotal reports of negative health effects from noise on people who live very close to wind turbines. Peer-reviewed research has generally not supported these claims.

The United States Air Force and Navy have expressed concern that siting large wind turbines near bases "will negatively impact radar to the point that air traffic controllers will lose the location of aircraft".

Before 2019, many wind turbine blades had been made of fiberglass with designs that only provided a service lifetime of 10 to 20 years. Given the available technology, as of February 2018 there was no market for recycling these old blades, and they were commonly disposed in landfills. Because blades are designed to be hollow, they take up large volume compared to their mass. Landfill operators have therefore started requiring operators to crush the blades before they can be landfilled.

New words and phrases

aerodynamics   *n.*   空气力学

anemometer   *n.*   风力计

arbitrage   *n.*   套利

baseload   *n.*   基本负载

biogas   *n.*   沼气

capacity factor   容量因素

dispatchable   *n.*   可调度的

doubly fed machine   双馈机

energy storage   储能

gearbox   *n.*   变速箱

geothermal   *adj.*   地热的

grid energy storage   电网储能

habitat   *n.*   栖息地

horizental   *adj.*   水平的

hydroelectric power   水力发电

hydrostorage   *n.*   水能储量

incandescent light bulb 白炽灯泡

induction generator 异步发电机

intermittent *adj.* 间歇性的

nacelle *n.* 吊篮;机舱

offshore farm 海上风电场

operating reserve 运行备用;营业准备

predictability *n.* 可预测性

pumped-storage plant 抽水储能电站

reactive power 无功功率

rotation *n.* 旋转

rotor blade 旋转叶片

rotor hub 转子轮毂

stall *n.* 摊位,小隔间

turbine *n.* 涡轮

variable renewable energy 变量可再生资源

variablity *n.* 可变性

vertical *adj.* 垂直的

wind farm 风力发电厂

 Exercises

Ⅰ. Warming-up questions

Directions：Give brief answers to the following questions.

1. What are the basic technologies in the conversion of energy from wind into electricity?

2. What kind of wind farms can be built and in what ways they are different?

3. Discuss the factors that influence the capability of wind power generation.

Ⅱ. Technical terms

Directions：Please give the Chinese or English equivalents of the following terms.

1. wind power generation   2. offshore wind farm

3. rotor blade

4. vertical axis

5. capacity ratio

6. reactive power

7. 空气力学

8. 白炽灯泡

9. 双馈机

10. 转子轮毂

11. 运行备用

12. 抽水储能电站

Ⅲ. Blank filling

Directions：Complete the following sentences by choosing words or phrases given below.

variable	intermittent	address	penetration	specification

1. "Why do you want to take a nice continuous process and turn it into an _____ one by coupling it with solar energy?" he asks.

2. This simple function takes in any _____ and displays it to the screen.

3. The granule enlarges, becomes glassy and transparent, and resists the _____ of various chemical substances.

4. This is important information for you, as an administrator, because you control the buffer pool size _____ and you can decide what size to specify.

5. Senior researchers in the field have made the public policy case for a Global Project on Artificial Photosynthesis to _____ critical energy security and environmental sustainability issues.

Ⅳ. Translation

Directions：Translate the following passages from English into Chinese or from Chinese into English.

Passage 1

Wind farms consist of many individual wind turbines, which are connected to the electric power transmission network. Onshore wind is an inexpensive source of electric power, competitive with or in many places cheaper than coal or gas plants. Onshore wind farms also have an impact on the landscape, as typically they need to be spread over more land than other power stations and need to be built in wild and rural areas, which can lead to "industrialization of the countryside" and habitat loss. Offshore wind is steadier and stronger than on land and offshore farms have less visual impact, but construction and maintenance costs are higher. Small onshore wind farms can feed some energy into the grid or provide electric power to isolated off-grid locations.

Passage 2

风能是一种间歇性能源,既不能直接产电,也不能按需调度。它还提供了可变功率,这种功率每年都是一致的,但在较短的时间范围内变化很大。因此,它必须与其他能源或存储器一起使用,才能提供稳定可靠的能源供给。随着风电在一个地区所占比例的增加,需要更多的常规能源(如化石燃料和核能)来支持,电网可能需要升级。如有可调度的能源、足够的水能、充足的容量、按地理分布的涡轮机、可向邻近地区输出和输入电力、能源存储情况理想或在风力发电量较低时可减少需求,在许多情况下都可以克服这些问题。天气预报使得电力网能为生产中发生的可预测变化做好准备。

 Additional reading

Unit 4
Solar Power Generation

Solar power generation is the conversion of energy from sunlight into electricity, either directly using photovoltaic (PV), indirectly using concentrated solar power, or a combination. Concentrated solar power systems use lenses or mirrors and tracking systems to focus a large area of sunlight into a small beam. Photovoltaic cells convert light into an electric current using the photovoltaic effect.

Photovoltaic was initially solely used as a source of electricity for small and medium-sized applications, from the calculator powered by a single solar cell to remote homes powered by an off-grid rooftop PV system. Commercial concentrated solar power plants were first developed in the 1980s. The 392 MW Ivanpah installation is the largest concentrating solar power plant in the world, located in the Mojave Desert of California.

As the cost of solar electricity has fallen, the number of grid-connected solar PV systems has grown into the millions and utility-scale photovoltaic power stations with hundreds of megawatts are being built. Solar PV is rapidly becoming an inexpensive, low-carbon technology to harness renewable energy from the Sun. The current largest photovoltaic power station in the world is the 850 MW Longyangxia Dam Solar Park, in Qinghai, China.

The International Energy Agency projected in 2014 that under its "high renewables" scenario, by 2050, solar photovoltaic and concentrated solar power would contribute about 16 and 11 percent, respectively, of the worldwide electricity consumption, and solar would be the world's largest source of electricity. Most solar

installations would be in China and India. In 2017, solar power provided 1.7% of total worldwide electricity production, growing at 35% per annul. As of 2018, the unsubsidised levelised cost of electricity for utility scale solar power is around $43/MWh.

1. The composition and equipment of solar power generation

Many industrialized nations have installed significant solar power capacity into their grids to supplement or provide an alternative to conventional energy sources while an increasing number of less developed nations have turned to solar to reduce dependence on expensive imported fuels. Long distance transmission allows remote renewable energy resources to displace fossil fuel consumption. Solar power plants use one of two technologies.

Photovoltaic systems use solar panels, either on rooftops or in ground-mounted solar farms, converting sunlight directly into electric power.

Concentrated solar power (CSP, also known as "concentrated solar thermal") plants use solar thermal energy to make steam, that is thereafter converted into electricity by a turbine.

1.1 Photovoltaic cells

A solar cell, or photovoltaic cell, is a device that converts light into electric current using the photovoltaic effect. The first solar cell was constructed by Charles Fritts in the 1880s. The German industrialist Ernst Werner von Siemens was among those who recognized the importance of this discovery. In 1931, the German engineer Bruno Lange developed a photo cell using silver selenide in place of copper oxide, although the prototype selenium cells converted less than 1% of incident light into electricity. Following the work of Russell Ohl in the 1940s, researchers Gerald Pearson, Calvin Fuller and Daryl Chapin created the silicon solar cell in 1954. These early solar cells cost $286/watt and reached efficiency of 4.5%-6%. In 1957, Mohamed M. Atalla developed the process of silicon surface passivation by thermal oxidation at Bell Labs. The surface passivation process has since been critical to solar cell efficiency.

The array of a photovoltaic power system, or PV system, produces direct current (DC) power which fluctuates with the sunlight's intensity. For practical use this usually requires conversion to certain desired voltages or alternating current (AC), through the use of inverters. Multiple solar cells are connected inside modules. Modules are wired together to form arrays, then tied to an inverter, which produces power at the desired voltage, and for AC, the desired frequency / phase.

Many residential PV systems are connected to the grid wherever available, especially in developed countries with large markets. In these grid-connected PV systems (see Fig. 4.1), use of energy storage is optional. In certain applications such as satellites, lighthouses, or in developing countries, batteries or additional power generators are often added as back-ups. Such stand-alone power systems permit operations at night and at other times of limited sunlight.

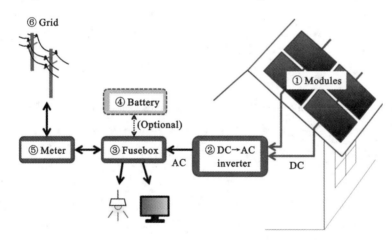

Figure 4.1　Schematics of a grid-connected residential PV power system

Source: https://en.wikipedia.org/wiki/Solar_power.

1.2　Concentrated solar power

Concentrated solar power, also called "concentrated solar thermal", uses lenses or mirrors and tracking systems to concentrate sunlight (see Fig. 4.2), then uses the resulting heat to generate electricity from conventional steam-driven turbines.

A wide range of concentrating technologies exists: among the best known are the parabolic trough, the compact linear Fresnel reflector, the Stirling dish and the solar power tower. Various techniques are used to track the sun and focus light. In all of these systems a working fluid is heated by the concentrated sunlight, and is then used

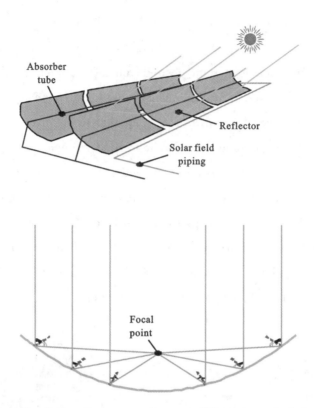

Figure 4. 2　A parabolic collector concentrates sunlight onto a tube in its focal point

Source：https：//en. wikipedia. org/wiki/Solar_power.

for power generation or energy storage. Thermal storage efficiently allows up to 24-hour electricity generation.

A parabolic trough consists of a linear parabolic reflector that concentrates light onto a receiver positioned along the reflector's focal line. The receiver is a tube positioned along the focal points of the linear parabolic mirror and is filled with a working fluid. The reflector is made to follow the sun during daylight hours by tracking along a single axis. Parabolic trough systems provide the best land-use factor of any solar technology. The SEGS plants in California and Acciona's Nevada Solar One near Boulder City, Nevada are representatives of this technology.

Compact linear fresnel reflectors are CSP-plants which use many thin mirror strips instead of parabolic mirrors to concentrate sunlight onto two tubes with working fluid. This has the advantage that flat mirrors can be used which are much cheaper than parabolic mirrors, and that more reflectors can be placed in the same amount of space, allowing more of the available sunlight to be used. Concentrating linear fresnel reflectors can be used in either large or more compact plants.

The stirling solar dish combines a parabolic concentrating dish with a stirling engine which normally drives an electric generator. The advantages of stirling solar over photovoltaic cells are higher efficiency of converting sunlight into electricity and longer lifetime. Parabolic dish systems give the highest efficiency among CSP technologies. The 50 kW Big Dish in Canberra, Australia is an example of this technology.

A solar power tower uses an array of tracking reflectors (heliostats) to concentrate light on a central receiver atop a tower. Power towers can achieve higher (thermal-to-electricity conversion) efficiency than linear tracking CSP schemes and better energy storage capability than dish stirling technologies. The PS10 Solar Power Plant and PS20 Solar Power Plant are examples of this technology.

1.3   Hybrid systems

A hybrid system combines PV/CPV and CSP with one another or with other forms of generation such as diesel, wind and biogas. The combined form of generation may enable the system to modulate power output as a function of demand or at least reduce the fluctuating nature of solar power and the consumption of non renewable fuel. Hybrid systems are most often found on islands.

1) CPV/CSP system

A novel solar CPV/CSP hybrid system has been proposed, combining concentrator photovoltaic with the non-PV technology of concentrated solar power, or also known as concentrated solar thermal.

2) ISCC system

The Hassi R' Mel power station in Algeria, is an example of combining CSP with a gas turbine, where a 25-megawatt CSP-parabolic trough array supplements a much larger 130 MW combined cycle gas turbine plant. Another example is the Yazd power station in Iran.

3) PVT system

Hybrid PVT, also known as photovoltaic thermal hybrid solar collectors convert solar radiation into thermal and electrical energy. Such a system combines a solar (PV) module with a solar thermal collector in a complementary way.

4) CPVT system

A concentrated photovoltaic thermal hybrid (CPVT) system is similar to a PVT system. It uses concentrated photovoltaic (CPV) instead of conventional PV technology, and combines it with a solar thermal collector.

5) PV diesel system

It combines a photovoltaic system with a diesel generator. Combinations with other renewables are possible and include wind turbines.

6) PV-thermoelectric system

Thermoelectric, or "thermovoltaic" devices convert a temperature difference between dissimilar materials into an electric current. Solar cells use only the high frequency part of the radiation, while the low frequency heat energy is wasted. Several patents about the use of thermoelectric devices in tandem with solar cells have been filed.

The idea is to increase the efficiency of the combined solar / thermoelectric system to convert the solar radiation into useful electricity.

2. The history and future of solar power generation

2. 1 Early days

The early development of solar technologies starting in the 1860s was driven by an expectation that coal would soon become scarce. Charles Fritts installed the world's first rooftop photovoltaic solar array, using 1%-efficient selenium cells, on a New York City roof in 1884. However, development of solar technologies stagnated in the early 20th century in the face of the increasing availability, economy, and utility of coal and petroleum. In 1974 it was estimated that only six private homes in all of North America were entirely heated or cooled by functional solar power systems. The 1973 oil embargo and 1979 energy crisis caused a reorganization of energy policies around the world and brought renewed attention to developing solar technologies. Deployment strategies focused on incentive programs such as the Federal Photovoltaic Utilization Program in the US and the Sunshine Program in Japan. Other efforts included the formation of research facilities in the United States (SERI, now NREL), Japan (NEDO), and Germany (Fraunhofer ISE). Between 1970 and 1983 installations of photovoltaic systems grew rapidly, but falling oil prices in the early 1980s moderated the growth of photovoltaic from 1984 to 1996.

2.2　Mid-1990s to early 2010s

In the mid-1990s development of both, residential and commercial rooftop solar as well as utility-scale photovoltaic power stations began to accelerate again due to supply issues with oil and natural gas, global warming concerns, and the improving economic position of PV relative to other energy technologies. In the early 2000s, the adoption of feed-in tariffs — a policy mechanism, that gives renewables priority on the grid and defines a fixed price for the generated electricity — led to a high level of investment security and to a soaring number of PV deployments in Europe.

2.3　Current status

For several years, worldwide growth of solar PV was driven by European deployment, but has since shifted to Asia, especially China and Japan, and to a growing number of countries and regions all over the world, including, but not limited to, Australia, Canada, Chile, India, Israel, Mexico, South Africa, South Korea, Thailand, and the United States.

Worldwide growth of photovoltaic has averaged 40% per year from 2000 to 2013 and total installed capacity reached 303 GW at the end of 2016 with China having the most cumulative installations (78 GW) and Honduras having the highest theoretical percentage of annual electricity usage which could be generated by solar PV (12.5%). The largest manufacturers are located in China.

Concentrated solar power also started to grow rapidly, increasing its capacity nearly tenfold from 2004 to 2013, albeit from a lower level and involving fewer countries than solar PV. As of the end of 2013, worldwide cumulative CSP-capacity reached 3,425 MW.

2.4　Forecasts

In 2010, the International Energy Agency predicted that global solar PV capacity could reach 3,000 GW or 11% of projected global electricity generation by 2050 — enough to generate 4,500 TWh of electricity. Four years later, in 2014, the agency projected that, under its "high renewables" scenario, solar power could

supply 27% of global electricity generation by 2050 (16% from PV and 11% from CSP).

3. The benefits and price of solar power generation

3.1　Cost

The typical cost factors for solar power include the costs of the modules, the frame to hold them, wiring, inverters, labor cost, any land that might be required, the grid connection, maintenance and the solar insulation that location will receive. Adjusting for inflation, it cost $96 per watt for a solar module in the mid-1970s. Process improvements and a very large boost in production have brought that figure down to 68 cents per watt in February 2016, according to data from Bloomberg New Energy Finance. Palo Alto California signed a wholesale purchase agreement in 2016 that secured solar power for 3.7 cents per kilowatt-hour. And in sunny Dubai large-scale solar generated electricity sold in 2016 for just 2.99 cents per kilowatt-hour — "competitive with any form of fossil-based electricity — and cheaper than most".

Photovoltaic systems use no fuel, and modules typically last 25 to 40 years. Thus, capital costs make up most of the cost of solar power. Operations and maintenance costs for new utility-scale solar plants in the US are estimated to be 9 percent of the cost of photovoltaic electricity, and 17 percent of the cost of solar thermal electricity. Governments have created various financial incentives to encourage the use of solar power, such as feed-in tariff programs. Also, renewable portfolio standards impose a government mandate that utilities generate or acquire a certain percentage of renewable power regardless of increased energy procurement costs. In most states, RPS goals can be achieved by any combination of solar, wind, biomass, landfill gas, ocean, geothermal, municipal solid waste, hydroelectric, hydrogen, or fuel cell technologies.

3.2　Levelized cost of electricity

The PV industry has adopted levelized cost of electricity (LCOE) as the unit of cost. The electrical energy generated is sold in units of kilowatt-hours (kWh). As a rule of thumb, and depending on the local insolation, 1 watt-peak of installed solar PV capacity generates about 1 to 2 kWh of electricity per year. This corresponds to a capacity factor of around 10%-20%. The product of the local cost of electricity and the insolation determines the break even point (BEP) for solar power. The International Conference on Solar Photovoltaic Investments, organized by EPIA, has estimated that PV systems will pay back their investors in 8 to 12 years. As a result, since 2006 it has been economical for investors to install photovoltaic for free in return for a long term power purchase agreement. Fifty percent of commercial systems in the United States were installed in this manner in 2007 and over 90% by 2009.

Shi Zhengrong has said that, as of 2012, unsubsidised solar power is already competitive with fossil fuels in India, Hawaii, Italy and Spain. He said, "we are at a tipping point. No longer are renewable power sources like solar and wind a luxury of the rich. They are now starting to compete in the real world without subsidies." "Solar power will be able to compete without subsidies against conventional power sources in half the world by 2015."

3.3　Current installation prices

In its 2014 edition of the *Technology Roadmap: Solar Photovoltaic Energy* report, the International Energy Agency (IEA) published prices for residential, commercial and utility-scale PV systems for eight major markets as of 2013. However, DOE's Sunshot Initiative has reported much lower US installation prices. In 2014, prices continued to decline. The Sunshot Initiative modeled US system prices to be in the range of $1.80 to $3.29 per watt. Other sources identify similar price ranges of $1.70 to $3.50 for the different market segments in the US, and in the highly penetrated German market, prices for residential and small commercial rooftop systems of up to 100 kW declined to $1.36 per watt by the end of 2014. In 2015, Deutsche Bank estimated costs for small residential rooftop systems in the US around $2.90 per watt. Costs for utility-scale systems in China and India were estimated as low as $1.00 per watt.

3.4　Grid parity

Grid parity, the point at which the cost of photovoltaic electricity is equal to or cheaper than the price of grid power, is more easily achieved in areas with abundant

sun and high costs for electricity such as in California and Japan. In 2008, the levelized cost of electricity for solar PV was $0.25/kWh or less in most of the OECD countries. By late 2011, the fully loaded cost was predicted to fall below $0.15/kWh for most of the OECD and to reach $0.10/kWh in sunnier regions. These cost levels are driving three emerging trends: vertical integration of the supply chain, origination of power purchase agreements (PPAs) by solar power companies, and unexpected risk for traditional power generation companies, grid operators and wind turbine manufacturers.

Grid parity was first reached in Spain in 2013, Hawaii and other islands that otherwise use fossil fuel (diesel fuel) to produce electricity, and most of the US is expected to reach grid parity by 2015.

In 2007, General Electric's Chief Engineer predicted grid parity without subsidies in sunny parts of the United States by around 2015; other companies predicted an earlier date: the cost of solar power will be below grid parity for more than half of residential customers and 10% of commercial customers in the OECD, as long as grid electricity prices do not decrease through 2010.

3.5　Productivity by location

The productivity of solar power in a region depends on solar irradiance, which varies through the day and is influenced by latitude and climate.

The locations with highest annual solar irradiance lie in the arid tropics and subtropics. Deserts lying in low latitudes usually have few clouds, and can receive sunshine for more than ten hours a day. These hot deserts form the Global Sun Belt circling the world. This belt consists of extensive swathes of land in Northern Africa, Southern Africa, Southwest Asia, Middle East, and Australia, as well as the much smaller deserts of North and South America. Africa's eastern Sahara Desert, also known as the Libyan Desert, has been observed to be the sunniest place on Earth according to NASA.

3.6　Self consumption

In cases of self consumption of the solar energy, the payback time is calculated based on how much electricity is not purchased from the grid. For example, in

Germany, with electricity prices of €0.25/kWh and insolation of 900 kWh/kWp, one kWp will save €225 per year, and with an installation cost of €1700/kWp the system cost will be returned in less than seven years. However, in many cases, the patterns of generation and consumption do not coincide, and some or all of the energy is fed back into the grid. The electricity is sold, and at other times when energy is taken from the grid, electricity is bought. The relative costs and prices obtained affect the economics. In many markets, the price paid for sold PV electricity is significantly lower than the price of bought electricity, which incentivizes self consumption. Moreover, separate self consumption incentives have been used in e. g. Germany and Italy. Grid interaction regulation has also included limitations of grid feed-in in some regions in Germany with high amounts of installed PV capacity. By increasing self consumption, the grid feed-in can be limited without curtailment, which wastes electricity.

A good match between generation and consumption is key for high self consumption, and should be considered when deciding where to install solar power and how to dimension the installation. The match can be improved with batteries or controllable electricity consumption. However, batteries are expensive and profitability may require provision of other services from them besides self consumption increase. Hot water storage tanks with electric heating with heat pumps or resistance heaters can provide low-cost storage for self consumption of solar power. Shiftable loads, such as dishwashers, tumble dryers and washing machines, can provide controllable consumption with only a limited effect on the users, but their effect on self consumption of solar power may be limited.

4. The application of solar power generation

The overwhelming majority of electricity produced worldwide is used immediately, since storage is usually more expensive and because traditional generators can adapt to demand. Both solar power and wind power are variable renewable energy, meaning that all available output must be taken whenever it is available by moving through transmission lines to where it can be used now. Since solar energy is not available at night, storing its energy is potentially an important issue particularly in off-grid and for future 100% renewable energy scenarios to have continuous electricity availability.

Solar electricity is inherently variable and predictable by time of day, location, and seasons. In addition, solar is intermittent due to day / night cycles and unpredictable weather. How much of a special challenge solar power is in any given electric utility varies significantly. In a summer peak utility, solar is well matched to daytime cooling demands. In winter peak utilities, solar displaces other forms of generation, reducing their capacity factors.

In an electricity system without grid energy storage, generation from stored fuels (coal, biomass, natural gas, nuclear) must go up and down in reaction to the rise and fall of solar electricity. While hydroelectric and natural gas plants can quickly respond to changes in load, coal, biomass and nuclear plants usually take considerable time to respond to load and can only be scheduled to follow the predictable variation. Depending on local circumstances, beyond about 20%-40% of total generation, grid-connected intermittent sources like solar tend to require investment in some combination of grid interconnections, energy storage (see Fig. 4.3) or demand side management. Integrating large amounts of solar power with existing generation equipment has caused issues in some cases. For example, in Germany, California and Hawaii, electricity prices have been known to go negative when solar is generating a lot of power, displacing existing baseload generation contracts.

Figure 4.3　Thermal energy storage

Source: https://en.wikipedia.org/wiki/Solar_power.

Conventional hydroelectricity works very well in conjunction with solar power;

water can be held back or released from a reservoir as required. Where a suitable river is not available, pumped-storage hydroelectricity (see Fig. 4.4) uses solar power to pump water to a high reservoir on sunny days, then the energy is recovered at night and in bad weather by releasing water via a hydroelectric plant to a low reservoir where the cycle can begin again. This cycle can lose 20% of the energy to round trip inefficiencies, this plus the construction costs add to the expense of implementing high levels of solar power.

Figure 4.4　Pumped-storage hydroelectricity (PSH)

Source: https://en.wikipedia.org/wiki/Solar_power.

Concentrated solar power plants may use thermal storage to store solar energy, such as in high-temperature molten salts. These salts are an effective storage medium because they are low-cost, have a high specific heat capacity, and can deliver heat at temperatures compatible with conventional power systems. This method of energy storage is used, for example, by the Solar Two power station, allowing it to store 1.44 TJ in its 68 m^3 storage tank, enough to provide full output for close to 39 hours, with an efficiency of about 99%.

In stand-alone PV systems batteries are traditionally used to store excess electricity. With grid-connected photovoltaic power system, excess electricity can be sent to the electrical grid. Net metering and feed-in tariff programs give these systems a credit for the electricity they produce. This credit offsets electricity provided from the

grid when the system cannot meet demand, effectively trading with the grid instead of storing excess electricity. Credits are normally rolled over from month to month and any remaining surplus settled annually. When wind and solar are a small fraction of the grid power, other generation techniques can adjust their output appropriately, but as these forms of variable power grow, additional balance on the grid is needed. As prices are rapidly declining, PV systems increasingly use rechargeable batteries to store a surplus to be later used at night. Batteries used for grid-storage stabilize the electrical grid by leveling out peak loads usually for several minutes, and in rare cases for hours. In the future, less expensive batteries could play an important role on the electrical grid, as they can charge during periods when generation exceeds demand and feed their stored energy into the grid when demand is higher than generation.

Although not permitted under the US National Electric Code, it is technically possible to have a "plug and play" PV microinverter. A recent review article found that careful system design would enable such systems to meet all technical, though not all safety requirements. There are several companies selling plug and play solar systems available on the web, but there is a concern that if people install their own it will reduce the enormous employment advantage solar has over fossil fuels.

Common battery technologies used in today's home PV systems include, the valve regulated lead-acid battery — a modified version of the conventional lead-acid battery, nickel-cadmium and lithium-ion batteries. Lead-acid batteries are currently the predominant technology used in small-scale, residential PV systems, due to their high reliability, low self discharge and investment and maintenance costs, despite shorter lifetime and lower energy density. Lithium-ion batteries have the potential to replace lead-acid batteries in the near future, as they are being intensively developed and lower prices are expected due to economies of scale provided by large production facilities such as the Gigafactory 1. In addition, the Li-ion batteries of plug-in electric cars may serve as a future storage devices in a vehicle-to-grid system. Since most vehicles are parked an average of 95% of the time, their batteries could be used to let electricity flow from the car to the power lines and back. Other rechargeable batteries used for distributed PV systems include, sodium-sulfur and vanadium redox batteries, two prominent types of a molten salt and a flow battery, respectively.

The combination of wind and solar PV has the advantage that the two sources complement each other because the peak operating times for each system occur at different times of the day and year. The power generation of such solar hybrid power systems is therefore more constant and fluctuates less than each of the two component subsystems. Solar power is seasonal, particularly in northern / southern climates, away from the equator, suggesting a need for long term seasonal storage in a medium such as hydrogen or pumped hydroelectric. The Institute for Solar Energy Supply Technology of the University of Kassel pilot-tested a combined power plant linking solar, wind, biogas and pumped-storage hydroelectricity to provide load-following power from renewable sources.

Research is also undertaken in this field of artificial photosynthesis. It involves the use of nanotechnology to store solar electromagnetic energy in chemical bonds, by splitting water to produce hydrogen fuel or then combining with carbon dioxide to make biopolymers such as methanol. Many large national and regional research projects on artificial photosynthesis are now trying to develop techniques integrating improved light capture, quantum coherence methods of electron transfer and cheap catalytic materials that operate under a variety of atmospheric conditions. Senior researchers in the field have made the public policy case for a Global Project on Artificial Photosynthesis to address critical energy security and environmental sustainability issues.

5. Environmental impacts

Unlike fossil fuel based technologies, solar power does not lead to any harmful emissions during operation, but the production of the panels leads to some amount of pollution.

5.1   Greenhouse gases

The life-cycle greenhouse-gas emissions of solar power are in the range of 22 to 46 gram (g) per kilowatt-hour (kWh) depending on if solar thermal or solar PV is being analyzed, respectively. With this potentially being decreased to 15 g/kWh in the future. For comparison (of weighted averages), a combined cycle gas-fired power plant emits some 400-599 g/kWh, an oil-fired power plant 893 g/kWh, a coal-fired power plant 915-994 g/kWh or with carbon capture and storage some 200 g/kWh, and a

geothermal high-temp. power plant 91-122 g/kWh. The life cycle emission intensity of hydro, wind and nuclear power are lower than solar's as of 2011 as published by the IPCC, and discussed in the article *Life-cycle Greenhouse-gas Emissions of Energy Sources*. Similar to all energy sources were their total life cycle emissions primarily lay in the construction and transportation phase, the switch to low carbon power in the manufacturing and transportation of solar devices would further reduce carbon emissions. BP Solar owns two factories built by Solarex (one in Maryland, the other in Virginia) in which all of the energy used to manufacture solar panels is produced by solar panels. A 1-kilowatt system eliminates the burning of approximately 170 pounds of coal, 300 pounds of carbon dioxide from being released into the atmosphere, and saves up to 400 litres of water consumption monthly.

The US National Renewable Energy Laboratory (NREL), in harmonizing the disparate estimates of life-cycle GHG emissions for solar PV, found that the most critical parameter was the solar insolation of the site: GHG emissions factors for PV solar are inversely proportional to insolation. For a site with insolation of 1700 kWh/m^2/year, typical of southern Europe, NREL researchers estimated GHG emissions of 45 g CO_2e/kWh. Using the same assumptions, at Phoenix, USA, with insolation of 2400 kWh/m^2/year, the GHG emissions factor would be reduced to 32 g of CO_2e/kWh.

The New Zealand Parliamentary Commissioner for the Environment found that the solar PV would have little impact on the country's greenhouse gas emissions. The country already generates 80 percent of its electricity from renewable resources (primarily hydroelectricity and geothermal) and national electricity usage peaks on winter evenings whereas solar generation peaks on summer afternoons, meaning a large uptake of solar PV would end up displacing other renewable generators before fossil-fueled power plants.

5.2 Energy payback

The energy payback time (EPBT) of a power generating system is the time required to generate as much energy as is consumed during production and lifetime operation of the system. Due to improving production technologies the payback time has been decreasing constantly since the introduction of PV systems in the energy market.

In 2000 the energy payback time of PV systems was estimated as 8 to 11 years and in 2006 this was estimated to be 1. 5 to 3. 5 years for crystalline silicon PV systems and 1-1. 5 years for thin film technologies. These figures fell to 0. 75-3. 5 years in 2013, with an average of about 2 years for crystalline silicon PV and CIS systems.

Another economic measure, closely related to the energy payback time, is the energy returned on energy invested (EROEI) or energy return on investment (EROI), which is the ratio of electricity generated divided by the energy required to build and maintain the equipment. (This is not the same as the economic return on investment (ROI), which varies according to local energy prices, subsidies available and metering techniques.) With expected lifetimes of 30 years, the EROEI of PV systems are in the range of 10 to 30, thus generating enough energy over their lifetimes to reproduce themselves many times (6-31 reproductions) depending on what type of material, balance of system (BOS), and the geographic location of the system.

5.3   Water use

Solar power includes plants with among the lowest water consumption per unit of electricity (photovoltaic), and also power plants with among the highest water consumption (concentrating solar power with wet-cooling systems).

Photovoltaic power plants use very little water for operations. Life-cycle water consumption for utility-scale operations is estimated to be 45 litres per megawatt-hour for flat-panel PV solar. Only wind power, which consumes essentially no water during operations, has a lower water consumption intensity.

Concentrating solar power plants with wet-cooling systems, on the other hand, have the highest water-consumption intensities of any conventional type of electric power plant; only fossil-fuel plants with carbon-capture and storage may have higher water intensities. A 2013 study comparing various sources of electricity found that the median water consumption during operations of concentrating solar power plants with wet cooling was 3. 1 cubic metres per megawatt-hour (810 US gal/MWh) for power tower plants and 3. 4 m^3/MWh (890 US gal/MWh) for trough plants. This was higher than the operational water consumption (with cooling towers) for nuclear at 2. 7 m^3/MWh (720 US gal/MWh), coal at 2. 0 m^3/MWh (530 US gal/MWh), or natural gas at 0. 79 m^3/MWh (210 US gal/MWh). A 2011 study by the National

Renewable Energy Laboratory came to similar conclusions: for power plants with cooling towers, water consumption during operations was 3.27 m^3/MWh (865 US gal/MWh) for CSP trough, 2.98 m^3/MWh (786 US gal/MWh) for CSP tower, 2.60 m^3/MWh (687 US gal/MWh) for coal, 2.54 m^3/MWh (672 US gal/MWh) for nuclear, and 0.75 m^3/MWh (198 US gal/MWh) for natural gas. The Solar Energy Industries Association noted that the Nevada Solar One trough CSP plant consumes 3.2 m^3/MWh (850 US gal/MWh). The issue of water consumption is heightened because CSP plants are often located in arid environments where water is scarce.

In 2007, the US Congress directed the Department of Energy to report on ways to reduce water consumption by CSP. The subsequent report noted that dry cooling technology was available that, although more expensive to build and operate, could reduce water consumption by CSP by 91% to 95%. A hybrid wet/dry cooling system could reduce water consumption by 32% to 58%. A 2015 report by NREL noted that of the 24 operating CSP power plants in the US, 4 used dry cooling systems. The four dry-cooled systems were the three power plants at the Ivanpah Solar Power Facility near Barstow, California, and the Genesis Solar Energy Project in Riverside County, California. Of 15 CSP projects under construction or development in the US as of March 2015, 6 were wet systems, 7 were dry systems, 1 hybrid, and 1 unspecified.

Although many older thermoelectric power plants with once-through cooling or cooling ponds use more water than CSP, meaning that more water passes through their systems, most of the cooling water returns to the water body available for other uses, and they consume less water by evaporation. For instance, the median coal power plant in the US with once-through cooling uses 138 m^3/MWh (36,350 US gal/MWh), but only 0.95 m^3/MWh (250 US gal/MWh) (less than one percent) is lost through evaporation. Since the 1970s, the majority of US power plants have used recirculating systems such as cooling towers rather than once-through systems.

5.4 Other issues

One issue that has often raised concerns is the use of cadmium (Cd), a toxic heavy metal that has the tendency to accumulate in ecological food chains. It is used as semiconductor component in CdTe solar cells and as a buffer layer for certain CIGS cells in the form of cadmium sulfide. The amount of cadmium used in thin-film solar cells is relatively small (5-10 g/m^2) and with proper recycling and emission control techniques in place the cadmium emissions from module production can be almost zero. Current PV technologies lead to cadmium emissions of 0. 3-0. 9 microgram/kWh over the whole life-cycle. Most of these emissions arise through the use of coal power for the manufacturing of the modules, and coal and lignite combustion leads to much higher emissions of cadmium. Life-cycle cadmium emissions from coal is 3. 1 microgram/kWh, lignite 6. 2, and natural gas 0. 2 microgram/kWh.

In a life-cycle analysis it has been noted, that if electricity produced by photovoltaic panels were used to manufacture the modules instead of electricity from burning coal, cadmium emissions from coal power usage in the manufacturing process could be entirely eliminated.

In the case of crystalline silicon modules, the solder material, that joins together the copper strings of the cells, contains about 36 percent of lead (Pb). Moreover, the paste used for screen printing front and back contacts contains traces of Pb and sometimes Cd as well. It is estimated that about 1,000 metric tonnes of Pb have been used for 100 gigawatts of c-Si solar modules. However, there is no fundamental need for lead in the solder alloy.

Some media sources have reported that concentrated solar power plants have injured or killed large numbers of birds due to intense heat from the concentrated sunrays. This adverse effect does not apply to PV solar power plants, and some of the claims may have been overstated or exaggerated.

A 2014-published life-cycle analysis of land use for various sources of electricity concluded that the large-scale implementation of solar and wind potentially reduces pollution-related environmental impacts. The study found that the land-use footprint, given in square meter-years per megawatt-hour (m^2a/MWh), was lowest for wind, natural gas and rooftop PV, with 0. 26, 0. 49 and 0. 59, respectively, and followed by utility-scale solar PV with 7. 9. For CSP, the footprint was 9 and 14, using parabolic troughs and solar towers, respectively. The largest footprint had coal-fired power plants with 18 m^2a/MWh.

6. Emerging technologies in solar power generation

6.1 Concentrator photovoltaic

Concentrator photovoltaic systems employ sunlight concentrated onto photovoltaic surfaces for the purpose of electrical power production. Contrary to conventional photovoltaic systems, it uses lenses and curved mirrors to focus sunlight onto small, but highly efficient, multi-junction solar cells. Solar concentrators of all varieties may be used, and these are often mounted on a solar tracker in order to keep the focal point upon the cell as the sun moves across the sky (see Fig. 4.5). Luminescent solar concentrators (when combined with a PV-solar cell) can also be regarded as a CPV system. Concentrated photovoltaic is useful as it can improve efficiency of PV-solar panels drastically.

Figure 4.5　CPV modules on dual axis solar trackers in Golmud, China

Source: https://en.wikipedia.org/wiki/Solar_power.

In addition, most solar panels on spacecraft are also made of high efficient multi-junction photovoltaic cells to derive electricity from sunlight when operating in the inner solar system.

6.2　Floatovoltaics

Floatovoltaics are an emerging form of PV systems that float on the surface of irrigation canals, water reservoirs, quarry lakes, and tailing ponds. Several systems exist in France, India, Japan, Korea, the United Kingdom and the United States. These systems reduce the need of valuable land area, save drinking water that would otherwise be lost through evaporation, and show a higher efficiency of solar energy conversion, as the panels are kept at a cooler temperature than they would be on land. Although not floating, other dual-use facilities with solar power include fisheries.

 New words and phrases

compact linear Fresnel reflector　紧凑的线性菲涅尔反射器

concentrated solar power　集中式太阳能发电

deployment　*n.*　调度;部署

diesel generator　柴油发电机

feed-in tariff　（对可再生的）能源补助

grid-connected PV　并网的光伏系统

hybrid system　混合动力系统

insolation　*n.*　日晒

inverter　*n.*　逆变器

landfill gas　垃圾填埋气（指垃圾分解而产生的一种含甲烷、二氧化碳等的混合气体）

latitude　*n.*　纬度

levelized cost of electricity　电力成本均衡

off-grid　*adj.*　离网的

parabolic collector　抛物面收集器

parabolic trough　抛物线形槽

photovoltaic power station　光伏发电站

PV   *n.*   光伏（photovoltaic）

pumped-storage hydroelectricity   抽水储能

renewable energy   *n.*   可再生能源

renewable portfolio standard   可再生能源投资组合标准

solar irradiance   日照度

solar panel   太阳能电池板

solar power   太阳能

solar power tower   太阳能发电塔

solar thermo energy   太阳能热能源

stand-alone power system   独立的电源系统

Stirling dish   蝶式斯特林

thermal energy storage   能量存储

thermoelectric system   热电系统

working fluid   工作流体

 Exercises

Ⅰ. Warming-up questions

Directions：Give brief answers to the following questions.

1. What are the basic technologies in the conversion of energy from sunlight into electricity？

2. How does photovoltaic system work in order to generate electric power？

3. What is concentrated solar power？

Ⅱ. Technical terms

Directions：Please give the Chinese or English equivalents of the following terms.

1. solar power   2. photovoltaic

3. concentrated solar power   4. electric current

5. renewable energy   6. solar panel

7. 光伏发电站   8. 电网储能

9. 热电系统　　　　　　　10. 垃圾填埋气

11. 能量存储　　　　　　　12. 人工光合作用

Ⅲ. Blank filling

Directions：Complete the following sentences by choosing words or phrases given below.

installation	conversion	capacity	renewable	reflector

1. Solar power is the _____ of energy from sunlight into electricity.

2. Solar PV is rapidly becoming an inexpensive, low-carbon technology to harness _____ energy from the sun.

3. You will see how to define this during the _____ of the sample service client application.

4. A solar power tower uses an array of tracking _____ to concentrate light on a central receiver atop a tower.

5. _____ does not equal electricity. For all those turbines to be worthwhile, the wind has to blow in specific places at specific strengths for specific periods of time.

Ⅳ. Translation

Directions：Translate the following passages from English into Chinese or from Chinese into English.

Passage 1

Solar power is the conversion of energy from sunlight into electricity, either directly using photovoltaic (PV), indirectly using concentrated solar power, or a combination. Concentrated solar power systems use lenses or mirrors and tracking systems to focus a large area of sunlight into a small beam. Photovoltaic cells convert light into an electric current using the photovoltaic effect.

Passage 2

光伏发电系统或 PV 系统的阵列会产生直流电（DC）功率,该功率会随太阳

光的强度而产生波动。对于实际使用,这通常需要通过使用逆变器转换为某些所需的电压或交流电(AC)。多个太阳能电池连接在模块内部。模块连接在一起形成阵列,然后连接到逆变器,该逆变器以所需的电压产生功率,而对于交流电,则以所需的频率/相位产生功率。

 Additional reading

Unit 5
Science of Electricity

1. Electrostatic phenomena in ancient times

The study of electrostatic phenomena in ancient China and Greece is one of the ancient civilizations in the world. There are many records of electromagnetic phenomena in ancient books. For example, in the late Western Han Dynasty's *Spring and Autumn Wei Kaoyi Post* (around 20 BC), it is recorded that the friction tortoise (a shell of marine reptile similar to a turtle) can attract small objects. During the Eastern Han Dynasty, Wang Chong further described this phenomenon in *Lun Heng*: "Dunmou (hawksbill turtle) made mustard and magnet led needle." Zhang Hua of the Western Jin Dynasty wrote in his book *the Chronicles of Natural History* that "when people comb their hair and take off their clothes, there are people who have light when combing and untying their hair, and there are also people who have strong voice", which recorded the phenomenon that the comb and the hair rub together and the coat and the underwear of different materials rub together and electricity.

Ancient Greece is the birthplace of Western electromagnetism. Some electromagnetic phenomena are recorded in ancient Greek documents. Plato (427 BC-347 BC) once mentioned that "the attraction of amber and magnet is an observed wonder". It shows that the ancient Greeks discovered the phenomenon that amber attracts small objects (see Fig. 5. 1) in more than 300 BC.

Figure 5.1　Electrostatic effect in ancient times

Source：http：//blog. sina. com. cn/s/blog_5f37717a0102wpxi. html.

2. The systematic study of electrostatic phenomena

2.1　Gilbert's first study on electrostatic phenomena

Gilbert was the first to study electromagnetic phenomena systematically. Gilbert (1544-1603) was a doctor who was the royal doctor of Queen Elizabeth Ⅰ. In addition to his medical work, he devoted himself to the study of magnetic phenomena and friction electrification. It is the first group of people to systematically study the electrical and magnetic phenomena through experiments, with many important findings. He first determined that the attraction of amber and magnet are two different phenomena. The magnet itself is attractive, and the amber is rubbed; the magnet can only attract magnetic objects, and the rubbed amber can attract any small objects. Gilbert had done many experiments of friction and electrification with various substances, and found that in addition to amber, diamond, sapphire, crystal, glass, sulfur, hard resin, mica, rock salt, and etc. can also attract small objects after friction. Gilbert called the object that can attract small objects after friction electric, which means "amber", which is the origin of the word "electricity" in the western language. In order to determine whether a substance is a charged body, he invented the first electroscope that can be used for experiments, which uses a very thin metal rod with the center balanced on a tip and can rotate freely. Because the rod is very light, when the charged object is near the rod after friction, the rod is attracted and rotates.

2.2   Grick's invention of friction motor

Otto Grick invented friction starting motor Otto Grick (1602-1686) is a versatile engineer, who was mayor of Magdeburg in Germany for 35 years. In 1654, he did the famous Magdeburg hemispheric experiment with his own invented ventilator. In 1660, the first friction starter which can produce a large amount of electric charge was invented, which created conditions for further study of electricity. Grick's friction motor (see Fig. 5.2) is to pierce a football sized sulfur ball along its diameter, insert it into the iron shaft, and horizontally install it on the seat frame so that the ball can rotate around the iron shaft. When rotating, place the dry palm on the ball, and the hand rubs against the ball, thus generating electricity. Grick did many interesting experiments with the friction motor. Through the experiment, we can observe the electric spark when the

Figure 5.2   Grick's friction motor

Source: https://www.sohu.com/a/239710769_100023715.

object is discharging. Later, Newton improved the friction starting motor by replacing sulfur ball with glass ball. In the future, some people continue to improve the friction motor. Using it to do all kinds of experiments can produce novel electrical phenomena, especially powerful sparks, and sparks from people, which can cause the world's surprise. In 1840's Germany, the whole society was interested in the phenomenon of electricity. Many people, out of curiosity, bought friction motors for experiments as entertainment. Greatly popularized the knowledge of electricity.

2.3 Gray's finding of the conduction of electricity

Gray found the conduction of electricity (see Fig. 5.3). Stephen Gray (1666-1736) was born in a handicraft family in England, and he was skilled in crafts. In his later years, he was very interested in electrical experiments and conducted three years of continuous research. The most important contribution was to find the conduction phenomenon of electricity, and to confirm that some objects were conductive and some were non-conductive. Gray uses different materials to study how far electricity can travel. He has done many experiments with sticks, twine, fishing rods, and etc., the longest one is 650 feet. In order to suspend the hemp rope used to transmit electricity in the experiment, he used silk and copper wire to hang the hemp rope. It was found that when copper wire was used, electricity could not be transmitted along the hemp rope. He guessed that electricity might have run away through copper wires and nails. After further experiments, he found that electricity is easier to conduct through metal than through silk. Therefore, objects that are easy to pass through (such as metal) are called conducting bodies, while objects that are difficult to pass through (such as silk wires) are called non conducting bodies. Gray also did an interesting experiment: a child was hoisted horizontally with several thick silk ropes, and the child's arm was touched with the rubbed electrified glass tube, so the child's hand and body could attract feathers and copper scraps. This shows that the human body is also a conductor.

Figure 5.3 Gray's experiment on the conduction of electricity

Source: http://story.kedo.gov.cn/c/2018-07-18/922406.shtml.

2.4   Duffy's finding of two kinds of electricity

French scientist Duffy (1698-1739) studied chemistry at the academy of sciences in Paris. Inspired by Gray's work, electrical research began in 1732. It turned out that there were two kinds of electricity.

From ancient times until Gilbert studied electromagnetic phenomena, people only knew about the phenomenon of electrical attraction. In 1629 the Italian scholar Carpio (1586-1650) discovered that friction-charged amber attracted small objects to it and then repelled them. Later, many scholars studied this phenomenon and put forward various hypotheses to explain it. It was settled by Duffy more than a hundred years later.

Duffy first studied frictional electrification and its conduction, and conducted many experiments. He came to the conclusion that all objects except metals and soft materials could generate electricity by friction. Conductors must be supported by an insulator to be charged; the charge of an object has nothing to do with color.

Duffy studied the phenomenon of electrical repulsion, in which he attracted gold foil by rubbing electrically charged glass tubes, and the gold foil was repelled when it touched the tube. He thought it was when the two came in contact that the glass tube transmitted the electricity to the gold foil and then repelled it. Duffy approached the gold foil that was repelled by the electrified glass bar with a woolly resin bar. To his great surprise, the gold foil was not repelled by the resin bar, but attracted. After much experimentation, Duffy finally determined that there were two kinds of electricity, one of which he called vitreous electricity (now known as positive electricity) and the other called resin-electricity (now known as negative electricity). The characteristic of these two kinds of electricity is that they repel each other and attract each other.

2.5   The invention of the Leiden bottle

Physicist Musschenbrock (1692-1761) of Leiden University in the Netherlands was trying to find a way to conserve electricity when he saw that it was difficult to make the electricity carried by charged bodies disappear quickly in the air. On one occasion, he suspended a gun barrel with silk thread and used it to receive electricity from a glass ball of friction motor; at one end of the barrel hang a brass wire. The lower end of the copper wire is put into a glass bottle with water in it. Musschenbrock let his assistant hold the glass bottle in one hand while he turned the motor vigorously. The assistant accidentally touched his other hand against the barrel of the gun, felt a strong electric shock, and shouted. Musschenbrock and his assistant switched places, letting the assistant shake the motor, his hand to hold the bottle, the other hand to touch the barrel of the gun, also received electric shock. This experiment shows that putting a charged body in a glass bottle can preserve electricity. Later people called this storage bottle the Leiden bottle (see Fig. 5.4). After several improvements, the inner and outer surfaces of Leiden bottles are coated with metal foil, a metal rod is inserted into

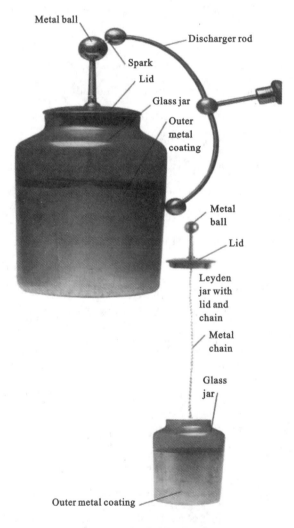

Figure 5.4   Leiden bottle

Source: http://www.changqingfoods.com/news.

the bottle cap, a metal ball is installed on the upper end of the rod, and a metal chain is used at the lower end to contact with the inner surface of the bottle. The bottle is filled with water, which increases the storage capacity of the bottle and can produce stronger electric shock. The electric spark produced during discharge can ignite gunpowder, hydrogen, and etc. The appearance of Leiden bottle provides a powerful means for further study of electrical phenomena.

The phenomenon of electric shock caused by the Leiden bottle aroused people's curiosity. The demonstration of electric shocks with Leiden bottles became a game of entertainment. One of the most remarkable performances was performed in front of Notre Dame cathedral by the French electrician Jean Noret. He asked the seven hundred monks to line up hand in hand, with the leader holding the Leiden bottle, and the tail holding the lead of the Leiden bottle. The audience was dumbfounded. The show showed the great power of electricity.

2.6   Franklin's kite experiment

It was in 1746 that Franklin (1706-1790) began his research into electricity after seeing the Leiden bottle experiment performed by Dr. Spencer from Scotland to the United States. Franklin was already 40 years old and successful in his career, becoming a celebrity. He was surprised and delighted to see Spence's electrical experiments. Franklin experimented enthusiastically with a glass tube given to his library by Collingson, a member of the royal society of London, and instructions for his experiments. Then he made experimental tools in his own glass factory. When he began to do experiments in electricity, he thought of lightning and sparks as the same thing, based on what he had observed, and assumed that lightning was caused by a large discharge of electricity from a charged cloud. He came up with the idea of using a kite to direct electricity from a thunderstorm cloud. He used a silk handkerchief to paste a kite, the kite was fitted with a pointed wire to attract the electricity in the cloud, the wire and the kite to fly with the string connected together, the string at the end of a ribbon and a metal key, the key as a conductor, in order to draw electricity, the kite held in the hands of the ribbon, to prevent the electricity through the body hurt. In July 1752, during a thunderstorm, Franklin, 46, took his son, 21, to the ranch and dropped the kite into the lightening clouds. They observed that the fibres of the twine stood on end, as though they had been electrically produced by friction. He held his finger close to the key, and an electric spark flashed across it. He charged the Leiden bottle and then discharged it, producing exactly the same effect as triboelectricity. This was the famous Franklin and Philadelphia kite experiment. It clearly proved that lightning is an electric discharge, which made a big step forward in the understanding of electricity. Later Franklin invented the lightning rod on this basis. After Franklin's kite experiment appeared, it was used everywhere to perform electrical experiments, and across the ocean to the United States.

2.7　The establishment of Coulomb's law

Coulomb (1736-1806) was a French engineer and scientist, who did many researches and published papers on civil engineering, mechanical mechanics and friction. He began to study electricity in 1785. Using his torsion scale to study the interaction between charged bodies, Coulomb's law was established. The precision and sensitivity of Coulomb's torsion scales made it possible to directly measure the weak electrostatic force between charges at different distances and to establish the inverse square law (see Fig. 5.5). The first international congress of electricity in 1881 decided to use the coulomb as a unit of electricity. The discovery of Coulomb's law made electricity enter the stage of quantitative science and laid a foundation for electrostatics.

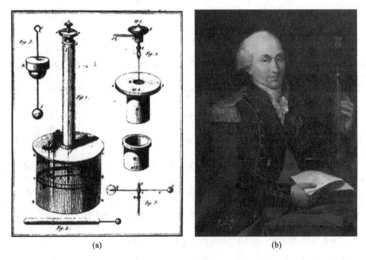

(a)   (b)

Figure 5.5　Coulomb and his experiment

Source: https://joyfulphysics.scholarnet.cn/? p=328.

2.8　Volta's invention of battery

Another important development in electricity in the late 18th century was the invention of the battery by the Italian physicist Volta. Prior to this, electrical experiments could only be carried out with Leiden bottles, the friction motors, and they could only provide a brief current. In 1780, the Italian anatomist Galvani happened to observe the twitching of a frog's leg in contact with metal. In further experiments, he found that when the two metals touched the tendons and muscles of the frog's leg, it twitched when the two metals touched each other.

After a careful study in 1792, Volta concluded that the twitch of the frog's leg was a sensitive response to electricity. The current is produced when two different metals are inserted into a solution to form a circuit, and the muscle provides the solution. With this in mind, in 1799 he built the first chemical battery capable of producing a continuous current (see Fig. 5.6), a series of cylinders of silver, zinc, and brine-soaked cardboard stacked in the same order, called a voltaic stack.

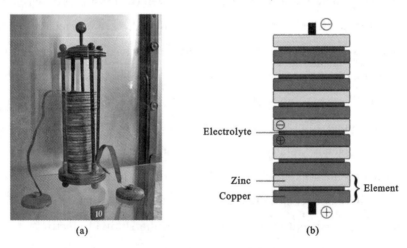

(a)　　　　　　　　　　　(b)

Figure 5.6　Voltaic battery

Source: https://pansci.asia/archives/49399.

Since then, various chemical power sources have flourished. In 1822 Seebeck further discovered that a weak and continuous current could also be obtained by connecting copper wire to a wire of another metal (bismuth) and maintaining a different temperature between the two joints. This was called the thermoelectric effect.

After the invention of chemical power, it was soon discovered that many unusual things could be done with it. In 1800 Carlyle and Nicholson used low-voltage currents

to break up water. In the same year Ritter successfully collected two gases from the electrolysis of water and electrolyzed copper from the solution of copper sulfate. In 1807, David used his huge battery to electrolyze potassium, sodium, calcium, magnesium and other metals. In 1811 he made a carbon arc out of a battery pack of 2,000 batteries. From the 1850s it became a powerful source of light for use in lighthouses, theatres and other places, and was gradually replaced by the incandescent lamp invented by Thomas Edison in the 1970s. Voltaic batteries also led to the development of electroplating, which was invented in 1839 by Siemens et al.

3. The scientific study on electromagnetic phenomena

3.1 Auster's discovery of the magnetic effect of current

Although Franklin had observed as early as 1750 that an electric discharge from a Leiden flask could magnetize a steel needle, and even earlier in 1640 that lightning had been observed to rotate the needle of a compass, by the early nineteenth century it was generally accepted in science that electricity and magnetism were two separate functions. Contrary to this traditional view, the Danish natural philosopher Auster accepted the philosophical idea of the unity of natural forces held by the German philosophers Immanuel Kant and Thomas Schelling, and believed that electricity and magnetism were somehow related. After years of research, he finally discovered in 1820 the magnetic effect of current (see Fig. 5. 7): when an electric current passes through a wire, it causes a magnetic needle near the wire to deflect. The discovery of the magnetic effect of current opened a new era in the study of electricity.

Auster's discovery first attracted the attention of French physicists, and in the same year some important achievements were made, such as Ampere's experiment on the equivalence of a current-carrying solenoid with a magnet. Arago on the magnetization of steel and iron under electric current; Biot and Savart's experiment on the force of a long direct current conductor on a magnetic pole; Ampere went on to do a series of ingenious experiments with the interaction of electric currents. The law of the interaction force between the current elements obtained from the analysis of these experiments is the basis for understanding the magnetic field generated by the current and the effect of the magnetic field on the current.

Figure 5. 7   Auster's magnetic field of current

Source: https://m. zol. com. cn/article/5720548. html.

3.2   Sturgeon's invention of electromagnet and further application

The discovery of the magnetic effect of current opened up a new field of electrical applications. In 1825, Sturgeon created the conditions for the widespread use of electricity by inventing the electromagnet. In 1833 Gauss and Weber produced the first simple single-wire telegraph. In 1837 Wheatstone and Morse independently invented the telegraph machine. Morse also invented a set of codes, which could be used to send messages by making dots and lines on moving paper.

In 1855 W. Thomson (Lord Kelvin) solved the problem of slow transmission of underwater cable signals, and in 1866 the Atlantic cable designed by Thomson was successfully laid. In 1854, the French telegrapher Charles Bourse conceived the idea of sending sound by electricity, but it didn't work. Later, in 1861, Rice's successful experiment went unnoticed. In 1861 Bell invented the telephone (see Fig. 5. 8), which is still used in modern times as a telephone receiver, while its transmitter was improved by Edison's carbon transmitter and Hughes's microphone.

Figure 5.8　Bell and the first telephone

Source: https://www.sohu.com/a/168568592_793537.

3.3　Ohm's discovery of circuit law

Shortly after the discovery of the magnetic effect of current, several different types of galvanometers were designed and made, which provided the conditions for Ohm to discover the circuit law (see Fig. 5.9). In 1826, inspired by Fourier's theory of heat conduction in solids, Ohm argued that the conduction of electricity is similar to the conduction of heat, and that a power source ACTS like a temperature difference in heat conduction. In order to determine the circuit law, he began to use the voltaic reactor as a power source to conduct experiments. Then he experimented with thermoelectric electromotive force, which was highly stable because of the constant temperature of the two contact points, and found that the strength of the current in the circuit was proportional to what he called the "test power" of the power supply, and the proportional coefficient was the resistance of the circuit.

Since the law of energy conservation had not been established at that time, the concept of electric power was ambiguous. It was not until 1848 that Kirchhoff examined the concepts of potential difference, electromotive force, and electric field intensity

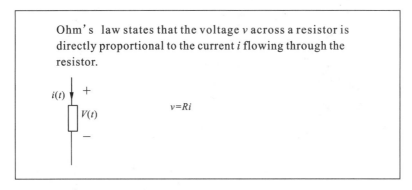

Figure 5.9   Circuit law

Source: https://www.mianfeiwendang.com/doc/3d437711a3fd9e41811ac58b/3.

from the perspective of energy, that ohmic theory was coordinated with the concept of electrostatics. On this basis, Kirchhoff solved the branch circuit problem.

3.4   Faraday's contribution to electromagnetics

3.4.1   The discovery of electromagnetic induction

Faraday, an outstanding British physicist, was engaged in experimental research on electromagnetic phenomena and made important contributions to the development of electromagnetics, the most important of which was the discovery of electromagnetic induction in 1831. Then he did a number of experiments to determine the law of electromagnetic induction(see Fig. 5.10). He found that when the magnetic flux in a closed coil changed, an induced electromotive force was generated in the coil. The size of the induced electromotive force depended on the rate of change of the magnetic flux over time. Later, Lenz gave a description of the direction of the induced current in 1834, and Noeman summarized their results to give the mathematical formula for the induced electromotive force.

Faraday made the first electric generator based on electromagnetic induction. In addition, he put "the electricity phenomenon is linked with other phenomena" widely studied, in 1833 successfully proved triboelectrification and voltaic cell to produce electricity are the same, was discovered in 1834's law of electrolysis, the effect, discovered in 1845, and explained the paramagnetic and diamagnetic material, he also studied the polarization phenomenon and static electric induction phenomenon in detail, and by experiment proves that the law of conservation of charge for the first time.

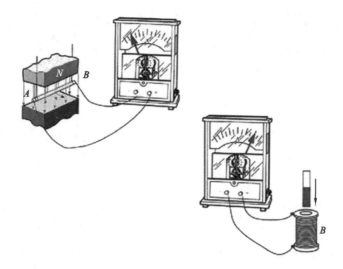

Figure 5. 10   Faraday's electromagnetic experiment

Source：http：//www. 51wendang. com/doc/27dbebf4e656d3a9103dbbfd/2.

The discovery of electromagnetic induction opens up a new prospect for the development and wide use of energy. In 1866 Siemens invented a practical self-excited motor. At the end of the 19th century electric power was transported over long distances; electric motors are widely used in production and transportation, which has greatly changed the face of industrial production.

3.4.2   Faraday's concept of field

The extensive study of electromagnetic phenomena led Faraday to develop his particular concept of "field". He thought: the line of force is material, it pervades all the space, and the different number of charge and the different magnetic plate connected separately; electricity and magnetism are transmitted not through the action of a distance in empty space, but through lines of force and magnetic force (see Fig. 5.11), which are essential components in the understanding of electromagnetic phenomena, and are even more valuable to study than the "sources" that produce or "converge" lines of force.

Faraday's abundant experimental research results and his novel concept of field prepared the conditions for the unified theory of electromagnetic phenomena. Noeman, Weber and other physicists have made many important contributions to the understanding of electromagnetic phenomena, but they did not succeed in establishing a unified theory by summarizing all the existing electrical knowledge since Coulomb from

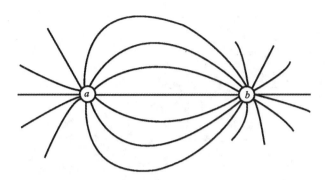

Figure 5. 11   Faraday's lines of force and magnetic force

Source: https://www. mianfeiwendang. com/doc/15a8e5463f16be6d8ab2b1c2.

the perspective of action at a distance. This work was done in the 1860s by the eminent British physicist John Maxwell.

3.5   Maxwell's theory of electromagnetic wave

Maxwell thought that a changing magnetic field excited a vortex electric field in the space around it. The changing electric field causes a change in the media's electrical displacement, which, like an electric current, stimulates a vortex magnetic field in the surrounding space. Maxwell explicitly expressed them in mathematical formulas, thus obtaining the general equations of electromagnetic fields — Maxwell's equations. Faraday's idea of line of force and electromagnetic transmission is fully embodied in it.

According to his equations, Maxwell further concluded that electromagnetic action propagates in the form of waves, and the propagation speed of electromagnetic wave in vacuum is equal to the ratio of electromagnetic and electrostatic units of electricity, and its value is the same as the propagation speed of light in vacuum, so Maxwell predicted that light is also an electromagnetic wave (see Fig. 5. 12).

In 1888, Hertz designed and made an electromagnetic wave source and an electromagnetic wave detector according to the oscillating nature of capacitor discharge. The electromagnetic wave was detected by experiments, and the wave velocity of the electromagnetic wave was measured. It was observed that the electromagnetic wave, like the light wave, had the property of polarization and could reflect, refract and focus. From then on, Maxwell's theory was gradually accepted by people.

Maxwell's electromagnetic theory has opened up a new field — the application and research of electromagnetic wave through the confirmation of Hertz electromagnetic

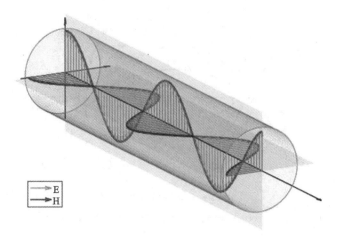

Figure 5.12   Maxwell's understanding on electromagnetic wave

Source: https://www.zhihu.com/question/68218231.

wave experiment. In 1895, radio signals were transmitted by Alexander Popov of Russia and Marconi of Italy. Marconi later modified the Hertz oscillator into a vertical antenna; Braun in Germany further divided the transmitter into two vibration lines, creating conditions for expanding the signal transmission range. In 1901 Marconi established the first transatlantic radio link. The invention of the electronic tube and its application in the circuit made it easy to transmit and receive electromagnetic waves, which promoted the development of radio technology and greatly changed human life.

3.6   Lorentz's theory of electrons

Lorentz's theory of electrons in 1896 applied Maxwell's equations to the microscopic field and reduced the electromagnetic properties of matter to the effect of electrons in atoms. In this way, it can not only explain the polarization, magnetization, electric conduction and other phenomena of matter, but also the absorption, scattering and dispersion of light. Moreover, the normal Zeeman effect of spectrum splitting in magnetic field has been successfully described. In addition, Lorentz derived the formula for the speed of light in the moving medium based on the theory of electrons, which further advanced Maxwell's theory.

The discovery of electrons combined electromagnetism with the theory of the structure of atoms and matter. Lorentz's theory of electrons reduced the macroscopic electromagnetic properties of matter to the effect of electrons in atoms, and explained the phenomena of electricity, magnetism and light in a unified way.

4. Einstein's special theory of relativity

In the theoretical system of Faraday, Maxwell and Lorentz, it is assumed that there is a particular medium called "ether", it is an electromagnetic wave load. Only in Ethernet reference frames, the speed of light in a vacuum is strictly directional independent, and Maxwell's equations and Lorentz force formulas are strictly established only in Ethernet reference frames. This means that the electromagnetic law does not conform to the relativity principle.

According to Maxwell's equations, the speed of light in a vacuum is a universal constant that is dependent only on the electrical permittivity and magnetic permeability of free space. This violates Galilean invariance, a long-standing cornerstone of classical mechanics. One way to reconcile the two theories (electromagnetism and classical mechanics) is to assume the existence of a luminiferous ether through which the light propagates. However, subsequent experimental efforts failed to detect the presence of the ether. After important contributions of Hendrik Lorentz and Henri Poincaré, in 1905, Albert Einstein solved the problem with the introduction of special relativity (see Fig. 5. 13), which replaced classical kinematics with a new theory of kinematics compatible with classical electromagnetism.

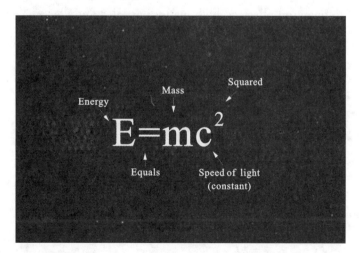

Figure 5. 13   Albert Einstein's special relativity

Source: http: //media. workercn. cn/sites/media/sxgrb/2017_07/18/GR0403. htm.

In addition, relativity theory implies that in moving frames of reference, a magnetic field transforms to a field with a nonzero electric component and conversely, a moving electric field transforms to a nonzero magnetic component, thus firmly showing that the phenomena are two sides of the same coin. The establishment of special relativity not only developed the electromagnetic theory, but also played a great role in the development of theoretical physics.

5. Quantum electromagnetic theory

Electricity, as a branch of classical physics, is well developed in terms of its fundamental principles, and can be used to explain electromagnetic phenomena in the macroscopic field.

In the 20th century, with the development of atomic physics, atomic nucleus physics and particle physics, human's understanding went deep into the micro field, in the interaction between charged particles and electromagnetic fields, the classical electromagnetic theory encountered difficulties. Although the classical theory has given some useful results, many phenomena cannot be explained by the classical theory. The limitation of the classical theory is that the description of charged particles ignores their volatility and the description of electromagnetic waves ignores their particle nature.

According to quantum physics, both matter particles and electromagnetic fields are both particle and volatile. Under the promotion of microscopic physics research, classical electromagnetic theory developed into quantum electromagnetic theory.

New words and phrases

amber　*n.*　琥珀

carbon arc　碳弧

charge　*v.*　充电

circuit　*n.*　电路

conduction　*n.*　传导

conductive　*adj.*　导电的

copper　*n.*　铜

cylinder　*n.*　油缸

diamagnetic　*adj.*　抗磁性的

diameter　*n.*　直径

discharge　*n.*　放电

dumbfounded　*adj.*　目瞪口呆的

electrical permittivity　介电常数

electromagnetism　*n.*　电磁学

electromotive　*adj.*　电动的;电动势的

electron　*n.*　电子

electroplating　*n.*　电镀

electrostatics　*n.*　静电学

flux　*n.*　通量

friction electrification　摩擦起电

galvanometer　*n.*　电流计

hemp rope　麻绳

induction　*n.*　（电磁）感应

insulator　*n.*　绝缘体

luminiferous ether　光以太

magnesium　*n.*　镁

magnetic permeability　磁导率

magnetite　*n.*　磁铁矿

mica　*n.*　云母

paramagnetic　*adj.*　顺磁性

potassium　*n.*　钾

repulsion　*n.*　排斥

resin　*n.*　树脂

resin electricity　树脂电

rod　*n.*　棒;枝条

sapphire　*n.*　蓝宝石

shaft　*n.*　轴

sodium　*n.*　钠

solenoid *n.* 螺线管

spark *n.* 电火花

special relativity 狭义相对论

static electricity 静电

sulfate *n.* 硫酸盐

sulfur *n.* 硫黄

thermo electric effect 热电效应

torsion *n.* 扭转

transmitter *n.* 发送器；变送器

triboelectricity *n.* 摩擦电

unified *adj.* 统一的

ventilator *n.* 通风设备

versatile *adj.* 多才多艺的

vitreous electricity 玻璃电荷

voltaic stack 伏打栈

 Exercises

I. Warming-up questions

Directions：Give brief answers to the following questions.

1. What are the observations of electricity in nature?

2. What is electric current? How was it discovered?

3. How does electricity and magnet interact with each other?

II. Technical terms

Directions：Please give the Chinese or English equivalents of the following terms.

1. static electricity

2. thermo electric effect

3. friction electrification

4. electric circuit

5. vitreous electricity

6. resin electricity

7. 光电效应

8. 光以太

9. 电流计   10. 抗磁性的

11. 狭义相对论   12. 量子电磁理论

III. Blank filling

Directions：Complete the following sentences by choosing words or phrases given below.

charge	friction	permeability	conductive	unified

1. He believed that _____ rendered amber magnetic, in contrast to minerals such as magnetite.

2. There was nothing in the brochure about having to drive it every day to _____ up the battery.

3. Salt water is more _____ than fresh water is.

4. A _____ theory of everything, including perhaps as many as 11 dimensions, would then be beckon.

5. Both types of paint are, though, permeable by oxygen and carbon dioxide, and that _____ means the red blood cells can still do their job.

IV. Translation

Directions：Translate the following passages from English into Chinese or from Chinese into English.

Passage 1

According to Maxwell's equations, the speed of light in a vacuum is a universal constant that is dependent only on the electrical permittivity and magnetic permeability of free space. This violates Galilean invariance, a long-standing cornerstone of classical mechanics. One way to reconcile the two theories (electromagnetism and classical mechanics) is to assume the existence of a luminiferous ether through which the light propagates. However, subsequent experimental efforts failed to detect the presence of the ether. After important contributions of Hendrik Lorentz and Henri Poincaré, in 1905, Albert Einstein solved the problem with the introduction of special relativity,

which replaced classical kinematics with a new theory of kinematics compatible with classical electromagnetism.

Passage 2

电学作为经典物理学的一个分支,就其基本原理而言,已发展得相当完善,它可用来说明宏观领域内的各种电磁现象。20世纪,随着原子物理学、原子核物理学和粒子物理学的发展,人类的认识深入到微观领域,在带电粒子与电磁场的相互作用问题上,经典电磁学理论遇到困难。虽然经典理论曾给出一些有用的结果,但是许多现象都是经典理论不能说明的。经典理论的局限性在于对带电粒子的描述忽略了其波动性方面,而对于电磁波的描述又忽略了其粒子性方面。按照量子物理的观点,无论是物质粒子或电磁场都既有粒子性,又具有波动性。在微观物理研究的推动下,经典电磁学理论发展为量子理论。

 Additional reading

Unit 6
Electric System

Electricity is an ideal energy form, which is convenient to deliver and use, and is clean without polluting our environment and atmosphere. It has developed rapidly and been widely used since it was discovered. The generation, delivery and consumption of electricity are realized in an integrated system which is called electric power system or power system.

1. Basic concepts of electric system

The electric power system is an electrical energy production and consumption system consisting of power plants, transmission and transformation lines, power supply and distribution stations, and power consumption.

Its function is to convert primary energy in the natural world into electrical energy through power generating devices (mainly including boilers, steam turbines, generators and auxiliary production systems of power plants, and etc.), and then supply electrical energy to various loads through transmission, transformation and distribution systems. The center, through various equipments, is converted into different forms of energy such as power, heat, and light to serve the local economy and people's lives. In order to achieve this function, the power system also has corresponding information and control systems at all links and at different levels. It measures, regulates, controls, protects, communicates and dispatches the production process of electrical energy to ensure that users get safe and high-quality electricity.

The main structure of the electric power system includes power sources (hydropower stations, thermal power plants, nuclear power stations and other power plants), substations (boost transformer substations, load center substations, and etc.), transmission and distribution lines and load centers (see Fig. 6. 1). The interconnection of power supply points can realize the power exchange and adjustment between regions, and improve the security and economy of power supply. The network formed by power transmission lines and transformers is usually called a power network. The information and control system of the power system consists of various detection equipment, communication equipment, safety protection devices, automatic control devices, and monitoring automation and dispatching automation systems. The structure of the power system should ensure the reasonable coordination of power production and consumption on the basis of advanced technology and equipment.

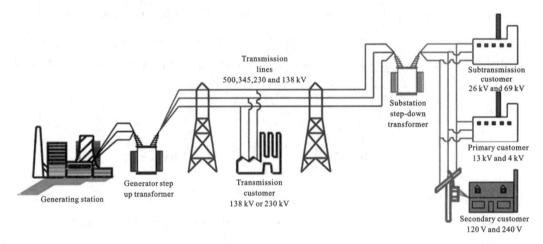

Figure 6. 1   Basic structure of the electric system

Source: https://www. webpages. uidaho. edu/sustainability/chapters/ch06/ch06-p3a. asp.

With the emergence of the power system, the highly efficient, pollution-free, easy-to-use, and easy-to-control electrical energy has been widely used, which has promoted changes in various fields of social production, pioneered the era of electricity, and has undergone a second technological revolution. The scale and technical level of the power system have become one of the signs of a country's economic development level.

2. The composition of electric system

2.1   Power plant

A power plant, also referred to as a power station and sometimes generating station or generating plant, is an industrial facility for the generation of electric power. Power stations are generally connected to an electrical grid.

Many power stations contain one or more generators, a rotating machine that converts mechanical power into three-phase electric power. The relative motion between a magnetic field and a conductor creates an electric current.

The energy source harnessed to turn the generator varies widely. Most power stations in the world burn fossil fuels such as coal, oil, and natural gas to generate electricity. Cleaner energy sources include nuclear power, and an increasing use of renewables such as solar, wind, wave and hydroelectric.

2.1.1   Thermal power station

In thermal power station (see Fig. 6.2), mechanical power is produced by a heat engine that transforms thermal energy, often from combustion of a fuel, into rotational energy. Most thermal power stations produce steam, so they are sometimes called steam power stations. Not all thermal energy can be transformed into mechanical power, according to the second law of thermodynamics; therefore, there is always heat loss to the environment. If this loss is employed as useful heat, for industrial processes or district heating, the power plant is referred to as a cogeneration power plant or CHP (combined heat-and-power) plant. In countries where district heating is common, there are dedicated heat plants called heat-only boiler stations. An important class of power stations in the Middle East uses by-product heat for the desalination of water.

The efficiency of a thermal power cycle is limited by the maximum working fluid temperature produced. The efficiency is not directly a function of the fuel used. For the same steam conditions, coal, nuclear and gas power plants all have the same theoretical efficiency. Overall, if a system is on constantly (base load) it will be more efficient than one that is used intermittently (peak load). Steam turbines generally operate at higher efficiency when operated at full capacity.

Figure 6.2   Thermal power station

Source: https://www.gruppo.acea.it/en/about-acea/our-plants/thermoelectric-power-plants.

Besides use of reject heat for process or district heating, one way to improve overall efficiency of a power plant is to combine two different thermodynamic cycles in a combined cycle plant. Most commonly, exhaust gases from a gas turbine are used to generate steam for a boiler and a steam turbine. The combination of a "top" cycle and a "bottom" cycle produces higher overall efficiency than either cycle can attain alone.

2.1.2   Classification

1) By heat source

Fossil-fuel power stations (see Fig. 6.3) may also use a steam turbine generator or in the case of natural gas-fired plants may use a combustion turbine. A coal-fired power station produces heat by burning coal in a steam boiler. The steam drives a steam turbine and generator, and then produces electricity. The waste products of combustion include ash, sulfur dioxide, nitrogen oxides, and carbon dioxide. Some of the gases can be removed from the waste stream to reduce pollution.

Nuclear power plants (see Fig. 6.4) use the heat generated in a nuclear reactor's core (by the fission process) to create steam which then operates a steam turbine and generator. About 20 percent of electric generation in the US is produced by nuclear power plants.

Figure 6.3   St. Clair Power Plant in Michigan, the United States

Source: en. wikipedia. org.

Figure 6.4   Ikata Nuclear Power Plant, Japan

Source: en. wikipedia. org.

Geothermal power plants (see Fig. 6.5) use steam extracted from hot underground rocks. These rocks are heated by the decay of radioactive material in the earth's crust.

Biomass-fuelled power plants may be fuelled by waste from sugar cane, municipal solid waste, landfill methane, or other forms of biomass.

Figure 6.5　Nesjavellir Geothermal Power Station, Iceland

Source: wikipediaen. wikipedia. org.

Waste heat from industrial processes is occasionally concentrated enough to use for power generation, usually in a steam boiler and turbine.

Solar thermal electric plants use sunlight to boil water and produce steam which turns the generator.

2) By prime mover

Steam turbine plants use the dynamic pressure generated by expanding steam to turn the blades of a turbine. Almost all large non-hydro plants use this system. About 90 percent of all electric power produced in the world is through the use of steam turbines.

Gas turbine plants use the dynamic pressure from flowing gases (air and combustion products) to directly operate the turbine. Natural-gas fueled (and oil fueled) combustion turbine plants can start rapidly and so are used to supply "peak" energy during periods of high demand, though at higher cost than base-loaded plants. These may be comparatively small units, and sometimes completely unmanned, being remotely operated. This type was pioneered by the UK, Princetown being the world's first, commissioned in 1959.

Combined cycle plants have both a gas turbine fired by natural gas, and a steam boiler and steam turbine which use the hot exhaust gas from the gas turbine to produce electricity. This greatly increases the overall efficiency of the plant, and many new base-load power plants are combined cycle plants fired by natural gas.

Internal combustion reciprocating engines are used to provide power for isolated communities and are frequently used for small cogeneration plants. Hospitals, office buildings, industrial plants, and other critical facilities also use them to provide backup power in case of a power outage. These are usually fuelled by diesel oil, heavy oil, natural gas, and landfill gas.

Micro-turbines, Stirling engines and internal combustion reciprocating engines are low-cost solutions for using opportunity fuels, such as landfill gas, digester gas from water treatment plants and waste gas from oil production.

2. 2　Substation

A substation is a part of an electrical generation, transmission, and distribution system. Substations transform voltages from high to low, or the reverse, or perform any of several other important functions. Between the generating station and consumer, electric power may flow through several substations at different voltage levels. A substation may include transformers to change voltage levels between high transmission voltages and lower distribution voltages, or at the interconnection of two different transmission voltages.

The word substation comes from the days before the distribution system became a grid. As central generation stations became larger, smaller generating plants were converted to distribution stations, receiving their energy supply from a larger plant instead of using their own generators. The first substations were connected to only one power station, where the generators were housed, and were subsidiaries of that power station.

Substations may be described by their voltage class, their applications within the power system, the method used to insulate most connections, and by the style and materials of the structures used. These categories are not disjointed; for example, to solve a particular problem, a transmission substation may include significant distribution functions.

2.2.1　Transmission substation

A transmission substation connects two or more transmission lines. The simplest case is where all transmission lines have the same voltage. In such cases, substation contains high-voltage switches that allow lines to be connected or isolated for fault clearance or maintenance. A transmission station may have transformers to convert between two transmission voltages, voltage control or power factor correction devices such as capacitors, reactors or static VAR compensators and equipment such as phase shifting transformers to control power flow between two adjacent power systems.

Transmission substations can range from simple to complex. A small "switching station" may be little more than a bus plus some circuit breakers. The largest transmission substations can cover a large area (several acres / hectares) with multiple voltage levels, many circuit breakers, and a large amount of protection and control equipment.

2.2.2　Distribution substation

A distribution substation transfers power from the transmission system to the distribution system of an area. It is uneconomical to directly connect electricity consumers to the main transmission network, unless they use large amounts of power, so the distribution station reduces voltage to a level suitable for local distribution.

The input for a distribution substation is typically at least two transmission or sub-transmission lines. Input voltage may be, for example, 115 kV, or whatever is common in the area. The output is a number of feeders. Distribution voltages are typically medium voltage, between 2.4 kV and 33 kV, depending on the size of the area served and the practices of the local utility. The feeders run along streets overhead (or underground, in some cases) and power the distribution transformers at or near the customer premises.

In addition to transforming voltage, distribution substations also isolate faults in either the transmission or distribution systems. Distribution substations are typically the points of voltage regulation, although on long distribution circuits (of several miles / kilometers), voltage regulation equipment may also be installed along the line.

The downtown areas of large cities feature complicated distribution substations, with high-voltage switching, and switching and backup systems on the low-voltage side.

More typical distribution substations have a switch, one transformer, and minimal facilities on the low-voltage side.

2.2.3　Collector substation

In distributed generation projects such as a wind farm or photovoltaic power station, a collector substation may be required (see Fig. 6. 6). It resembles a distribution substation although power flow is in the opposite direction, from many wind turbines or inverters up into the transmission grid. Usually for economy of construction the collector system operates around 35 kV, although some collector systems are 12 kV, and the collector substation steps up voltage to a transmission voltage for the grid. The collector substation can also provide power factor correction if it is needed, metering, and control of the wind farm. In some special cases a collector substation can also contain an high voltage direct current (HVDC) converter station.

Figure 6. 6　Substation in Russia

Source: https://en. wikipedia. org/wiki/Electrical_substation#/media/File: Switchgear_110_kV. jpg.

Collector substations also exist where multiple thermal or hydroelectric power plants of comparable output power are in proximity. Examples for such substations are Brauweiler in Germany and Hradec in the Czech Republic, where power is collected from nearby lignite-fired power plants. If no transformers are required for increasing the voltage to transmission level, the substation is a switching station.

2.2.4　Converter substation

Converter substations may be associated with HVDC converter plants, traction current, or interconnected non-synchronous networks. These stations contain power electronic devices to change the frequency of current, or else convert from alternating to direct current or the reverse. Formerly rotary converters changed frequency to interconnect two systems; nowadays such substations are rare.

2.2.5 Switching station

A switching station is a substation without transformers and operating only at a single voltage level. Switching stations are sometimes used as collector and distribution stations. Sometimes they are used for switching the current to back-up lines or for parallelizing circuits in case of failure. An example is the switching stations for the HVDC Inga-Shaba transmission line.

A switching station may also be known as a switchyard, and these are commonly located directly adjacent to or nearby a power station (see Fig. 6.7). In this case the generators from the power station supply their power into the yard onto the generator bus on one side of the yard, and the transmission lines take their power from a feeder bus on the other side of the yard.

Figure 6.7　Switchyard at Grand Coulee Dam, the United States

Source：wikiwand. com.

2.3　Transmission lines

Most power plants are far away from the load center. For example, coal resources in China are mainly concentrated in Shanxi, northern Shaanxi, Inner Mongolia, and western Henan. Hydraulic resources are mainly distributed in the southwest and northwest of China. The load centers are in the east and south, and the coastal power load accounts for 75% of the country. This requires the use of power lines as a channel for transmitting power, and the power lines that send power from the power plant to the load center are called transmission lines. There are currently two transmission methods, three-phase AC transmission and ultra-high voltage DC transmission, and three-phase

AC transmission is the main transmission method in China. The maximum voltage level of three-phase AC transmission lines in China is 1000 kV, and the DC transmission lines are ±800 kV.

Power lines are divided into overhead lines and cable lines. In overhead lines (see Fig. 6. 8), conductors are erected on line pole towers, and cable lines are generally laid underground. Because overhead line construction costs are much lower than cable lines, and overhead lines are easy to construct, maintain, and repair, so most of the lines in the power network use overhead lines, and cable lines are considered only when overhead lines cannot be used due to environmental restrictions, such as distribution networks in large cities.

Figure 6. 8   Overhead transmission lines

Source: https://pixabay.com/zh/photos/power-electricity-line-pylon-1549122/.

2. 4   Electric power distribution

Distribution is the final stage in the delivery of electric power; it carries electricity from the transmission system to individual consumers. Distribution substations connect to the transmission system and lower the transmission voltage to medium voltage ranging between 2 kV and 35 kV with the use of transformers. Primary distribution lines carry this medium voltage power to distribution transformers located near the customer's premises. Distribution transformers (see Fig. 6.9) again lower the voltage to the utilization voltage used by lighting, industrial equipment or household appliances. Often several customers are supplied from one transformer through secondary distribution lines. Commercial and residential customers are connected to the secondary distribution lines through service drops. Customers demanding a much larger amount of power may be connected directly to the primary distribution level or the subtransmission level.

Figure 6.9 Pole-mounted distribution transformers

Source: https://pixabay.com/zh/photos/electrician-worker-work-electricity-3771316/.

The transition from transmission to distribution happens in a power substation, which has the following functions:

(1) Circuit breakers and switches enable the substation to be disconnected from the transmission grid or for distribution lines to be disconnected.

(2) Transformers step down transmission voltages, 35 kV or more, down to primary distribution voltages. These are medium voltage circuits, usually 600-35,000 V.

(3) From the transformer, power goes to the busbar that can split the distribution power off in multiple directions. The bus distributes power to distribution lines, which fan out to customers.

Urban distribution is mainly underground, sometimes in common utility ducts. Rural distribution is mostly above ground with utility poles, and suburban distribution is a mix. Closer to the customer, a distribution transformer steps the primary distribution power down to a low-voltage secondary circuit, usually 120/240 V in the US for residential customers. The power comes to the customer via a service drop and an electricity meter. The final circuit in an urban system may be less than 15 metres, but may be over 91 metres for a rural customer.

2.5 Electric power users

The units that use electrical energy are called power users. The sum of the power required by all electrical equipment of a power consumer is called the electrical load. Electric power users can be divided into three categories according to the requirements for power supply reliability.

1) Class I users

Stopping the power supply to such loads will bring personal danger, damage to equipment, generate a large number of waste products, and disrupt production order for a long time, causing huge losses to the national economy or causing significant political impact. Class I users have reliability requirements for power supply.

Class I users should have more than two independent power sources, and the capacity of each power source should be guaranteed separately in this power. In the case of power supply, the user's power requirements can be met, ensuring that when any one of the power sources fails or is repaired, the power supply to the user will not be interrupted.

2) Class II users

Stopping the supply of power to such loads will cause a large amount of output reduction, and urban public utilities and people's lives will be affected (traffic lighting, large factories, and etc.). Class II users should set up dedicated power supply lines. When conditions permit, dual-circuit power supply can also be used, and priority should be given to ensuring their power supply when there is insufficient power supply.

3) Class III users

Class III users generally refers to users who will not cause serious consequences

after a short power outage, such as factory affiliated workshops, small towns, and small processing plants. One type of power supply can be set for Class Ⅲ users. When an accident occurs in the system and the power supply is insufficient, the load of Class Ⅲ users should be cut off first to ensure the power consumption of Class Ⅰ and Ⅱ users.

3. A brief history of China's electric power development

China's electric power system is gradually formed with the development of China's power industry. Its development can be divided into the following three stages.

3.1　Birth

At 7: 00 pm on July 26, 1882, 15 arc photoelectric lights connected in series on the 6. 4-km-long power line on the Bund lit (see Fig. 6. 10). Electricity for street lights came from China's first power plant. On that day, China's power industry started in Shanghai. However, the first power plant born on the Chinese land belonged to the Shanghai Electric Company founded by the British.

Figure 6. 10　Street lights from China's first power plant

Source: https: //baijiahao. baidu. com/s? id = 1607032241530812633&wfr = spider&for = pc.

In 1888, the first electric light in Beijing was lit in the Empress Dowager Cixi. Officials and wealthy households in Beijing successively used electric lights.

Before the 1911 Revolution, more than 20 cities in China had built new electric light factories, and correspondingly there were a small number of power supply lines. The country's total installed power generation capacity is only 27,000 kilowatts.

Initially, power generation and power supply were for lighting. Most power plants were called electric lamp factories, and electricity bills had to be calculated according to lamp holders. Electricity is a luxury for ordinary people. The power supply mode is only close-to-point point-to-point transmission, and there is no grid.

In 1912, in the context of the transient prosperity of national industries, government-run, commercial-run, and government-commercial joint-venture electric power enterprises were established in various places. Electricity began to be used more in mining, factories, transportation, planting (irrigation) and other industries.

In 1912, the first hydropower station in China, the Shilongba Hydropower Station in Kunming, Yunnan (see Fig. 6. 11), was put into operation. This was the beginning of China's hydropower business.

Figure 6. 11　The Shilongba Hydropower Station

Source: https://baijiahao. baidu. com/s? id=1607032241530812633&wfr=spider&for=pc.

In 1921, the 32-km-long Kunming Shilongba Hydropower Station transmission line was completed, and China had the first 10,000-volt transmission line.

In 1930, the National Government promulgated the "Electric Business Regulations", which gradually unified frequency and voltage. The frequency is 50 Hz, and the user voltage is 380 volts and 220 volts. This is the earliest standardization in China's electric power industry, and it is still used today.

Prior to 1949, industrially developed Shanghai had a large urban power grid. North China and Northeast China, which had been invaded by Japan, appeared China's first power grid, the Pingjintang Power Grid, and the only inter-provincial power grid at that time, the Northeast Power Grid.

In addition, the highest voltage levels elsewhere are 33 kV, 13.2 kV and 3.3 kV. The grid pattern is mainly urban isolated grid.

By 1949, China's power industry, which had gone through wars and tribulations, had a history of 67 years, but it was like a weak child-withered power plants, worn-out equipment, weak power grids, and difficult operation. In this year, the United States produced more than 60 times the annual power generation of China.

3.2 Newborn

On October 1, 1949, Chairman Mao pressed an electric button on the Tiananmen Gate, and the first five-star red flag in front of Tiananmen was raised by electricity (see Fig. 6.12). New China was born, and China's power industry also gained new vitality.

In 1953, China launched its first five-year plan. As the main infrastructure, the power industry has begun to develop on a large scale. In 1956, the first domestic 6,000-kilowatt thermal power unit was put into operation at Huainan Power Plant, and finally China can manufacture its own thermal power equipment.

The growth and extension of the power grid has always been accompanied by the needs of economic and social development and the construction and development of power sources.

In 1955, the first 110-kilovolt power transmission line designed and constructed in China, the Jingguan Line, was completed to cooperate with the power transmission of the Guanting Hydropower Station. This is nearly 50 years later than the same level abroad.

Figure 6.12   The first five-star red flag raised by electricity

Source: https://baijiahao.baidu.com/s? id=1607032241530812633&wfr=spider&for=pc.

In 1969, Liujiaxia Hydropower Station, the first million-kilowatt hydropower station designed, constructed and manufactured in China, was put into operation on the upper reaches of the Yellow River, the mother river of the Chinese. In 1972, the Liutianguan line of Liujiaxia Power Plant for hydropower delivery was put into operation. This is China's first 330-kV transmission line, which is still 20 years later than the world's equivalent transmission project.

We are still trying to catch up, the gap still exists, and it is gradually narrowing. In the 1960s and 1970s, in addition to the Northwest Power Grid, China's power grids were gradually interconnected through 220-kilovolt lines, with 220-kilovolt lines as the main grid, and provincial grids with provinces as the main power supply scope.

During this period, the development of electricity was mainly to serve the needs of infrastructure construction and industrial production. "Upstairs, downstairs, light phones." It was people's longing for a better future in that era.

3.3   Vitality

Reform and opening-up has injected a strong impetus into the economy and society, and the demand for electricity has become increasingly strong. The situation of "supplying more monks and less porridge" in electricity supply lasted for some years. Especially in the early days of reform and opening up, there was a sudden power outage during cooking, a sudden power outage in the classroom during self-study, and even a sudden power outage while working in the workshop... candles and flashlights are essential household items.

In response to power shortages, in the 1980s, the national power industry began to do everything possible to build electricity, build power plants, and generate more power. A number of large hydropower stations, pithead thermal power stations and nuclear power stations have been successively constructed.

At the end of 1988, the Gezhouba Water Conservancy Project, the first dam on the Yangtze River at that time (see Fig. 6. 13), was completed with an installed capacity of 2. 715 million kilowatts.

Figure 6. 13　Gezhouba Water Conservancy Project

Source: http://www.chinadaily.com.cn/m/hubei/gezhouba/2014-02/24/content_17166702.htm.

In 2009, the Three Gorges Hydropower Project was completed, with a total installed capacity of 22. 5 million kilowatts and an annual power generation capacity of nearly 100 billion kWh (see Fig. 6. 14). It is still the world's largest power station.

In December 1991, the Qinshan Nuclear Power Plant (see Fig. 6. 15) designed and built by China was successfully connected to the grid, achieving a breakthrough in nuclear power in mainland China.

Figure 6. 14   Three Gorges Hydropower Project

Source: https://www. chinadaily. com. cn/m/gezhouba/2014-02/24/content_17166777. htm.

Figure 6. 15   Qinshan Nuclear Power Plant

Source: alchetron. com.

In 1994, China introduced foreign technology to build the second nuclear power plant in the Mainland, the Daya Bay Nuclear Power Station (see Fig. 6. 16).

The rapid development of power supplies urgently requires higher voltage levels and stronger power grids. On December 22, 1981, China's first 500-kilovolt ultra-high-voltage transmission line, from Pingdingshan, Henan, to Wuchang, Hubei, was

Figure 6.16　Daya Bay Nuclear Power Station

Source：https：//www. chinadaily. com. cn/m/gezhouba/2014-02/25/content_17560113. htm.

completed. The two provinces of Hubei and Henan have strengthened the interconnection and expanded the power grid capacity. China became the 8th country in the world to have a 500-kV line.

The running pace has never stopped. In the 21st century, China's power industry has accumulated a lot.

In 2004, China became the country with the largest installed hydropower capacity in the world. In 2014, the country's hydropower generation capacity exceeded 1 trillion kWh for the first time, accounting for about one fifth of the country's total power generation.

As of the end of October 2015, China had the largest number of nuclear power plants under construction in the world, and also possessed independent intellectual property rights and advanced three-generation nuclear power technologies in the world. Large-scale power sources are accompanied by advanced large power grids.

In January 2009, the 1000 kV UHV AC test demonstration project in Southeast Jinan-Nanyang-Jingmen was completed. In July 2010, the Xiangjiaba-Shanghai ±800 kV UHV DC project was completed and put into operation. China's power grid has fully entered the era of UHV AC and DC hybrid power grids.

In 2011, the Qinghai-Tibet interconnection project was put into operation; in November 2014, the Sichuan-Tibet interconnection project was put into operation. With the exception of Taiwan Province, the grids in all provinces in China have realized AC and DC interconnection.

In 2014, China's UHV technology went abroad. In 2015, China's nuclear power technology went abroad. China Power has become a resounding name in the world.

More than just scale, China's electricity production, transmission, and consumption are all more "green". At the end of 2004, China's installed wind power capacity was only 740,000 kilowatts, and in 2011 it reached 62.36 million kilowatts, making China the world's largest wind power country. By 2017, the total installed capacity of photovoltaic power generation in China has exceeded 130 million kilowatts, and the newly added installed capacity has ranked first in the world for several consecutive years.

In 2017 and 2018, Qinghai Power Grid realized the use of clean energy for 7 consecutive days and 9 consecutive days. This is a new world record.

For more than 100 years, the 6.4-km street light line has become the world's largest power grid. Power surges across mountains and rivers into towns and villages, and the bloodline extends to millions of households.

 New words and phrases

adjacent   *adj.*   邻近的;毗连的

alternating   *adj.*   交变的;交互的

boost transformer substation   升压变电站

bus   *n.*   母线;总线;汇流条

busbar   *n.*   母线

capacitor   *n.*   电容器

cogeneration   *n.*   废热发电(利用废气热能发电)

combustion   *n.*   燃烧;氧化

combustion turbine   燃气轮机

converter station   换流站

desalination   *n.*   (海水的)脱盐;淡化

dispatching automation   调度自动化

distribution transformer   配电变压器;配电站

district heating　集中供热;区域供暖

dual-circuit　*n.*　双电路

extract　*v.*　提取,提炼

fault　*n.*　故障

fault clearance　故障排除

feeder　*n.*　馈电线

fission　*n.*　裂变;分裂

harness　*v.*　治理;利用

hydropower station　水电站

input voltage　输入电压

interconnected　*adj.*　连通的;有联系的

load center substation　负荷中心变电站

non-synchronous　*adj.*　不同步的

parallelize　*v.*　平行放置;使……平行于……

phase shifting　相移

power factor　功率因数;功率系数

power loss　功率损耗

prime mover　原动机

regulation　*n.*　调整

static VAR compensator　静止无功补偿器

steam turbine　汽轮机

step-down transformer　降压变压器

sulfur dioxide　二氧化硫

switching station　转换站

thermodynamic　*adj.*　热力学的;使用热动力的

traction current　牵引电流

transformer　*n.*　变压器

collector substation　集热变电站

transient　*adj.*　短暂的;瞬时的

transmission substation　输电变电站

voltage drop　电压降;电压凹陷

Exercises

Ⅰ. Warming-up questions

Directions：Give brief answers to the following questions.

1. What are the basic components of electric power system?

2. Why are overhead transmission lines more popular than underground lines?

3. What are the types of power plants and substations? Explain the production process of the power plant and the role of the substation.

Ⅱ. Technical terms

Directions：Please give the Chinese or English equivalents of the following terms.

1. power electronic   2. transformer

3. power plant   4. hydropower station

5. distribution substation   6. prime mover

7. 三相   8. 汽轮机

9. 电压降   10. 输入电压

11. 发电机   12. 故障排除

Ⅲ. Blank filling

Directions：Complete the following sentences by choosing words or phrases given below.

transmission	substation	utilization	synchronous	convert

1. Power generation is the first stage of the whole progress of the _____ of electric energy.

2. A distribution system connects all the individual loads to the transmission lines at _____ which performs voltage transformation.

3. Interconnection to neighboring power systems are usually formed at the _____ system level.

4. The electric power system is one of the tools for _____ and transporting energy.

5. Power system use _____ machines for generation of electricity.

IV. Translation

Directions: Translate the following passages from English into Chinese or from Chinese into English.

Passage 1

A thermodynamic cycle consists of a linked sequence of thermodynamic processes that involve transfer of heat and work into and out of the system, while varying pressure, temperature, and other state variables within the system, and that eventually returns the system to its initial state. In the process of passing through a cycle, the working fluid (system) may convert heat from a warm source into useful work, and dispose of the remaining heat to a cold sink, thereby acting as a heat engine. Conversely, the cycle may be reversed and use work to move heat from a cold source and transfer it to a warm sink thereby acting as a heat pump. At every point in the cycle, the system is in thermodynamic equilibrium, so the cycle is reversible.

Passage 2

在中国和印度的带领下,太阳能光伏发电迅速发展,太阳能将在 2040 年成为低碳发电的最大来源,届时可再生能源在总发电量中的比例将达到 40%。在欧盟国家,可再生能源占新增装机容量的 80%,由于陆上和海上风电的有力发展,预计 2030 年后不久,风力发电将成为主要的电力来源方式。政策会继续在全球范围内支持可再生能源发电,越来越多地通过竞争性拍卖而不是能源补贴来扶持这项产业,而数百万家庭、社区和企业直接投资于分布式太阳能光伏发电,从而推动电力行业的转型。在世界范围内的诸多能源中,电能仍将日益增长,到 2040 年其最终消费增长将达 40%,与石油在过去 25 年中所占的增长份额相同。

Additional reading

Unit 7
Electrical Equipment and Electricity Safety

The important role that electricity plays in our lives and production cannot be ignored. It brings us great convenience and has become an important energy source in our production and life. The most critical factor in a power plant that enables normal operation and transmission of electricity is electrical equipment. Electrical equipment is a collective name for generators, transformers, power lines, circuit breakers and other equipment in the power system.

1. Electrical equipments and their roles in the power system

1.1 Classification of electrical equipment

According to the requirements for the safety, high quality, reliability, and economic operation of the generation, transformation, transmission, distribution, and use of electrical energy in power plants and substations, the following electrical equipment are mainly available.

1.1.1 Primary equipment

Electrical equipment directly involved in the production, transmission, and distribution of electrical energy is called primary equipment, and it usually includes the following categories.

(1) Equipment for producing and converting electrical energy, such as generators and transformers.

(2) Switching devices that turn on or off the circuit, such as circuit breakers, disconnectors, fuses, contactors, and etc.

(3) Current-carrying conductors that connect related electrical equipment into circuits, such as bus cables.

(4) Equipment for converting electrical quantities, voltage transformers and current transformers.

(5) Electrical protective devices, such as reactors and arresters.

(6) Grounding device. There are working grounding, protective grounding and lightning protection grounding in the power system.

Generators, transformers, motors, and etc. belong to this category. The generator and the main transformer are the heart of the power station, referred to as the host and the main transformer.

1.1.2　Electrical secondary equipment

Electrical secondary equipment refers to the low-voltage electrical equipment required to monitor, control, adjust, and protect the work of the primary equipment, and to provide operating and maintenance personnel with operating conditions or production command signals, such as fuses, buttons, indicators, control switches, relays, control cables, meters, signaling equipment, automatic devices, and etc.

Secondary equipment does not directly participate in the production and distribution of electrical energy, but plays a very important role in ensuring the normal and orderly operation of the main equipment and the economic benefits of its operation.

2. Main electrical primary equipments in the power grid and their functions

2.1　Generator

Generator refers to mechanical equipment that converts other forms of energy into electrical energy (see Fig. 7.1). It is driven by water turbines, steam turbines, diesel engines or other power machinery. It converts the energy generated by water flow, gas flow, fuel combustion or nuclear fission into mechanical energy and transmits it to the generator. It is then converted into electricity by a generator.

Generators are widely used in industrial and agricultural production, national defense, technology and daily life. There are many forms of generators, but their working principles are based on the law of electromagnetic induction and the law of electromagnetic force. Therefore, the general principle of its structure is using appropriate magnetically conductive and conductive materials to form magnetic circuits and circuits that perform electromagnetic induction with each other to generate electromagnetic power and achieve the purpose of energy conversion.

Figure 7.1　A diesel generator

Source: https://pixabay.com/zh/photos/diesel-generators-generator-1462424/.

2.2　Transformer

Transformer is a kind of static electrical equipment that realizes electric energy transmission through electromagnetic coupling through the principle of electromagnetic induction. It is a device used to change an AC voltage (current) of a certain value into another voltage (current) of the same frequency or different values. It is mainly composed of an iron core and a winding wound on the core.

Transformer is one of the main equipments of power plants and substations. The role of the transformer is not only to increase the voltage to send electrical energy to the area where it is used, but also to reduce the voltage to the voltage used at all levels to meet the demand for electricity. In short, both step-up and step-down must be done by a transformer. In the process of transmitting electrical energy in the power system, voltage and power losses are bound to occur. When the same power is transmitted, voltage loss is inversely proportional to voltage, and power loss is inversely proportional to the square of the voltage. The transformer is used to increase the voltage and reduce the transmission loss.

2.3 High-voltage circuit breaker

High-voltage circuit breaker (also called high-voltage switch, see Fig. 7.2) can not only cut off or close the no-load current and load current in the high-voltage circuit, but also cut off the overload current and short-circuit current through the role of the relay protection device when the system fails within a specified time. It has a fairly complete arc extinguishing structure and sufficient interruption capacity.

Figure 7.2 High-voltage circuit breaker

Source: fashiontrendsnow. com.

High-voltage circuit breaker plays a controlling role in high-voltage circuits and is one of the important electrical components in high-voltage circuits. The circuit breaker is used to switch on or off the circuit during normal operation. In the case of a fault, the circuit is quickly disconnected under the action of the relay protection device. In special cases (such as when it is automatically reconnected to the fault line), the short-circuit current is reliably connected. The important roles are as follows.

(1) Control effect. According to the needs of power system operation, it can put some or all electrical equipment and some or all lines into or out of operation.

(2) Protection. When a part of the power system fails, it cooperates with protection devices and automatic devices to quickly remove the faulty part from the system, reduce the scope of power outages, prevent accidents from expanding, protect various electrical equipment in the system from damage, and ensure the system safe operation without faults.

2.4 Isolating switch

Isolating switch, also known as knife gate, is a high-voltage switchgear (see Fig. 7.3). Because it does not have a special arc extinguishing device, it cannot be used to cut off the load current and short-circuit current. It should be used in conjunction with the circuit breaker, and can only be operated when the circuit breaker is open. When the disconnector is opened, a visible break is formed between the dynamic and static contacts, and the insulation is reliable.

Isolating switch is mainly used in low-voltage terminal power distribution systems such as residential buildings and buildings in low-voltage equipment. It can play important roles as follows.

(1) After opening the gate, establish a reliable insulation gap, and separate the equipment or line that needs to be repaired from the power supply with an obvious disconnection point to ensure the safety of the maintenance personnel and equipment.

(2) Change the line according to the operation needs.

(3) It can be used for small currents in the opening and closing lines, such as the charging current of bushings, connectors, short cables, the capacitor current of the switching voltage equalizing capacitor, the circulating current when the double busbars are connected, and the excitation current of the voltage transformer.

(4) According to the specific situation of different structure types, it can be used to divide and close the no-load excitation current of a certain capacity transformer.

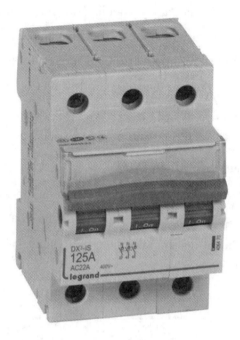

Figure 7. 3   The isolating switch

Source: https://www. legrand. com/ecatalogue/406470-isolating-switch-3p-400-v. html.

3. Electricity safety

This section focuses on the safe use of electricity, and introduces the relevant knowledge of human body electric shock, methods and safety appliances for safe electricity use, the causes and precautions of electric shock, first-aid methods of electric shock, electrical fire prevention, explosion prevention, and general knowledge of lightning protection.

3. 1   What is an electric shock?

Electric shock usually refers to tissue damage and dysfunction caused by direct human body contact with electric power or high voltage electricity passing through human body by way of air or other conductive media. Common electric shocks are direct contact with charged objects, making the human body part of the circuit, such as standing on the ground and touching the live line of the power supply, leakage of conductive objects that have a higher voltage to the ground, or direct contact with two power lines; the other is human body standing too close to the high-voltage line, causing an arc discharge of current through the human body, or near a high-voltage wire that falls on the ground, in an electric field formed by high-voltage electricity on the ground, the potential difference between the two feet of the human body (the so-called step voltage) making people get an electric shock.

There are three ways that people get an electric shock.

1) Single-phase electric shock

In a low-voltage power system, if a person stands on the ground and touches a live wire, it is a single-phase or single-wire shock, as shown in the Fig. 7.4. Human contacting leakage equipment housing also belongs to single-phase electric shock.

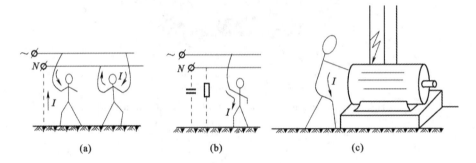

Figure 7.4   Single-phase electric shock

2) Two-phase shock

The electric shock caused by different parts of the body contacting the charged body of two-phase power source at the same time is called two-phase electric shock, as shown in the Fig. 7.5.

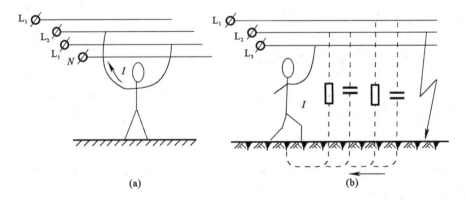

Figure 7.5   Two-phase electric shock

3）**Contact voltage and step voltage shock**

When the insulation of the electrical equipment with the grounding of the enclosure is damaged and the enclosure is electrified, or the conductor is broken and single-phase grounding fault occurs, the current from the enclosure of the equipment flows into the earth through the grounding wire, the grounding body (or through the grounding point through the downed conductor), diffuses to the surrounding area, and forms a strong electric field at and around the wiring connection point.

Contact voltage is the voltage at which a person standing on the ground touches the device housing. Step voltage is the voltage between the feet of a person standing on the ground near the device. (See Fig. 7.6.)

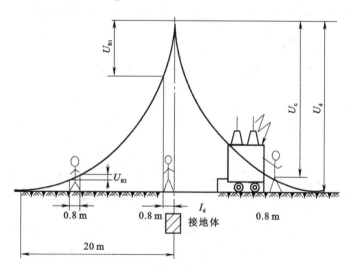

Figure 7.6   Contact voltage and step voltage shock

3.2   Causes of electric shock and its symptoms

3.2.1   Causes of electric shock

Statistics show that the main reasons for electric shock accidents are as follows.

(1) Lack of electrical safety knowledge. Fly kites near high-voltage lines, climb high-voltage poles to dig bird's nests; after low-voltage overhead lines are broken, keep picking up live wires by hand; touch live parts by hand with live wires at night; touch broken plastic covers by hand knife gate.

(2) Violation of operating procedures. Connect live lines or electrical equipment without taking the necessary safety measures; touch the damaged equipment or wires;

misplace the live equipment; connect the lighting fixtures live; repair the electric tools with live; move the electrical equipment with live; twist the light bulb with wet hands.

(3) The equipment is unqualified, and the safety distance is not enough; the ground resistance of the second wire and ground system is too large; the ground wire is unqualified or the ground wire is disconnected.

(4) The equipment is out of repair, the wind blows the line or the pole is not repaired in time; the bakelite of the rubber-covered knife gate is not changed in time; the motor wire is broken and the shell is charged for a long time; the porcelain bottle is damaged, the phase wire and the pull wire are shorted, and the equipment shell is charged.

(5) Other accidental causes, such as walking at night touching a live wire that has broken on the ground.

3.2.2 The symptoms of an electric shock

1) Electric shock

When the human body is exposed to electric current (see Fig 7.7), in mild cases the person immediately appears panic, stagnation, pale, the contracted site muscle contraction, and has dizziness, tachycardia, and general weakness. In severe cases, coma, persistent twitching, ventricular fibrillation, heartbeat, and respiratory arrest were observed. Some patients with severe electric shock, although the symptoms are not severe at the time, can suddenly worsen after 1 hour. Some patients have a very weak heartbeat and breathing after an electric shock, and even temporarily stop. They are in a state of "fake death". Therefore, they must be carefully identified and the rescue of patients with electric shock cannot be easily abandoned.

2) Electrothermal burns

Electric current burns more heavily at the entrance to the skin than at the exit. The burned skin was grayish-yellow scorched skin with a low depression in the center and no inflammatory reactions such as swelling and pain. But soft tissue burns on the current path are often more severe. After a large piece of soft tissue of the limb is electrically burned, chemia and necrosis often occur in the distant tissues. Increased plasma myosin and increased red blood cell membrane-induced hemoglobin can cause acute tubular necrotizing nephropathy.

(a)

(b)   (c)

Figure 7.7   Electric shock

3) Lightning damage

When a person is struck by lightning, heartbeat and breathing often stop immediately, with myocardial damage. Skin vasoconstriction has a network pattern, which is considered to be a characteristic of lightning damage. Other clinical manifestations are similar to high-voltage electrical injury.

3.3   How are electric shocks treated?

The specific operation of rescuing patients with electric shock is mainly determined by the degree of electric shock. For light patients, such as transient exposure to low voltage and weak current, only manifested as nervousness, pale, dull expression, heartbeat, and rapid breathing. Sensitive patients have a short-term syncope or coma, but can recover quickly. For this type of patients, as long as the power supply has been disconnected, no special first-aid measures are necessary, and they can recover after rest. However, they should closely observe the changes in vital signs, pay attention to the localized electric burns, and give corresponding symptomatic treatment. If the current intensity of the patient is large, the duration of the electric shock is long, the symptoms are severe, and even cardiac arrest and respiratory arrest occur, the following rescue principles should be followed to deal with it.

3.3.1   Rescue at the scene of an electric shock

The on-site treatment is against time, and the first task is to cut off the power.

According to the environment and conditions of the electric shock site, take the safest and fastest way to cut off the power or remove the electric shock person from the power source.

1) Turning off the power

If an electric shock occurs at home or near the switch, it is the easiest, safe and effective way to quickly turn off the power switch and open the main power switch.

2) Disconnecting the wire

Use a dry wooden stick, bamboo pole, and etc. to lift the wire away from the person who is electrocuted, and fix the wire to prevent others from getting an electric shock.

3) Cutting off the circuit

If the cut-off circuit is in the field or away from the power switch, especially on rainy days, when it is inconvenient to approach the electric shock to pick up the power cord, you can use insulated pliers or a dry wooden handle shovel, axe, knife 20 meters away waiting for the wire to be cut.

4) Pulling away the electric shock person

If the electric shock scene is far away from the switch or does not have the condition to turn off the power, the rescuer can stand on the dry board, holding the clothes with one hand to pull them away from the power, also can use dry wood stick, bamboo pole to wait to pick out electric wire from electric person's body (see Fig. 7.8).

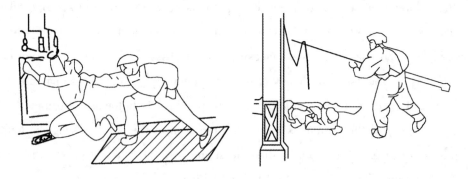

(a) Pull the electrocuted person away from the power supply　　(b) Remove the wires from the victim

Figure 7.8　Pull away the electric shock person

3.3.2　Treatment for different conditions

(1) The electrocuted person is still conscious, but feels dizzy, or palpitations, cold sweats, nausea, vomiting, and etc. occur, he should be allowed to lie still to reduce the burden of the heart.

(2) The electrocuted person is sometimes conscious, sometimes unconscious, should lie still, and call a doctor for treatment.

(3) The electrocuted person has no consciousness, breathing or heartbeat. Artificial respiration should be performed while a doctor is called in.

(4) The electrocuted person stops breathing, but the heartbeat still exists, and artificial respiration should be performed; if the heartbeat is stopped, and breathing is still present, the external chest compression method should be adopted; if both breathing and heartbeat stop, both artificial respiration and external chest compression should be performed.

3.3.3　Two first aid methods

1) Artificial respiration

Only for those who have stopped breathing, artificial respiration should be performed. The steps are as follows (see Fig. 7.9).

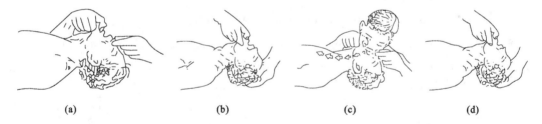

(a)　　　　　(b)　　　　　(c)　　　　　(d)

Figure 7.9　Artificial respiration process

① Make the eletrocuted person lie on his back first, unbutton the collar, scarf, tight clothes, etc. , remove mucus, blood, food, false teeth and other sundries in the mouth.

② Tilt the head back as far as possible, nose in the air and neck straight. The rescuer holds the eletrocuted person's nostrils with one hand and opens his mouth with the other. After inhaling deeply, the rescuer breathes deeply into his mouth, causing his chest to expand. After the rescue person breathe, relax the electric shock of the mouth and nose, so that it automatically exhale. Do this over and over again, blowing for 2 seconds, relaxing for 3 seconds, about 5 seconds a cycle.

③ When blowing, you should pinch your nostril and close it to your mouth to prevent air leakage. When relaxing, you should make him exhale automatically.

④ If the teeth of the eletrocuted person are closed and cannot be pried open, the method of blowing air from the mouth to the nose can be adopted.

⑤ The force of blowing on the weak and children should be slightly lighter to avoid alveoli rupture.

2) Cardiopulmonary resuscitation (CPR)

The brain needs a lot of oxygen. After the breathing and heartbeat stop, the brain will soon be short of oxygen, and half of the brain cells will be damaged in 4 minutes. If cardiopulmonary resuscitation is performed more than 5 minutes later, only a quarter of the population may survive.

① For those who have weak or irregular breathing, or even stop breathing, but the heartbeat is still present, they should immediately breathe artificially from mouth to mouth, or get supine chest compressions, prone pressure back-type artificial respiration, and those with conditions can use tracheal intubation balloon or ventilator to assist breathing.

② For patients with cardiac arrest and respiration, chest compressions should be performed immediately. For patients with ventricular fibrillation, non-synchronous DC defibrillation should be performed when conditions permit.

③ Those who have a heartbeat or respiratory arrest should be given cardiopulmonary resuscitation (CPR, see Fig. 7.10) immediately.

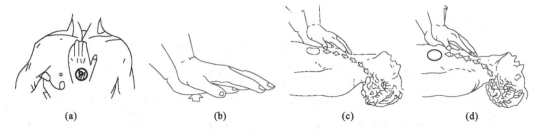

(a)   (b)   (c)   (d)

Figure 7.10 Cardiopulmonary resuscitation process

3.4 Precautions against electric shock

It's vitally important to take safety precautions when working with electricity. Safety must not be compromised and some ground rules need to be followed first. The basic guidelines regarding safe handling of electricity documented below will help you while working with electricity.

(1) Never mix water and electricity. Always keep electrical appliances away from water and moisture.

(2) Pay attention to what your appliances are telling you. When an appliance repeatedly trips a circuit breaker, blows a fuse, or gives you shocks, it's not just a coincidence. You need to take preventive measure and be careful about those signs.

(3) Install ground fault circuit interrupters (GFCI). In new construction homes, GFCI receptacles are a requirement anywhere that electrical outlets and water are in close proximity to one another. You should install them at the correct place to prevent severe electrical injuries.

(4) Make sure you're using the right size circuit breakers and fuses. If fuses and circuit breakers aren't the right size and wattage rating to match the specifications of their circuits, they're going to fail right when you most need them to perform. Read packages carefully when shopping for replacements. If you're not sure which size to buy, have an electrician take a look at your panel box and label it with the circuit breaker or fuse size needed (for easy future reference). And as long as you're making a trip to the hardware store, stock up with a few extra — you'll be happy to have them on hand when the next need arises.

(5) Protect kids with outlet covers. Outlet covers prevent babies and small children from sticking their fingers and other objects into unoccupied receptacles, protecting them against shock and electrocution. You can either use the plug-in type, or opt for special child safety wall plates, which feature built-in, retractable covers that automatically snap back into place when outlets aren't in use.

(6) Avoid cube taps and other outlet-stretching devices. Cubes taps — those little boxes that allow you to plug several appliances into a single outlet — may seem like a major convenience, but they can actually put you on the fast track to circuit overload, overheated wiring, and even fire. If you absolutely must use one, do the math before plugging in. Know the maximum power demand that the cube-tapped receptacle can handle, and be certain that the collective pull (power requirement) of the devices you're plugging into it doesn't exceed that rating.

(7) Replace missing or broken wall plates. They're not just there for the looks

— wall plates also protect your fingers from making contact with the electrical wiring behind them. Broken wall plates, or the absence of them altogether, can be especially dangerous in the dark — when trying to locate a switch by touch, you may end up being shocked or electrocuted if you miss the mark and end up hitting live wires instead.

(8) Keep electrically powered yard-care tools dry. Whether it's raining, just finished raining, or you've recently run the sprinklers, never attempt yard work with electrically powered tools in wet conditions. Protect yourself from shock and electrocution by keeping your electric hedge trimmer, weed whacker, and lawnmower safely unplugged and stowed away until precipitation has stopped, grass and foliage are dry, and puddles can be easily avoided.

(9) Match the light bulb's wattage rating to the lamp. Whenever choosing light bulbs to match with a lamp, be sure to consult that lamp's maximum wattage specifications (they're often printed right around the light bulb socket). Always opt for a light bulb with wattage that's equal to or less than the maximum wattage listed on the lamp — too strong a bulb can lead to overloaded lamp wiring, as well as fire.

(10) Be kind to your power cords. Take care to treat power cords gently — never nail or tightly tack them down, and regularly check to make sure that they're not pinched between or underneath furniture. Excessive pressure on power cables can damage insulation (exposing the conductor), or compress the conducting wire, which can lead to overheating and put you at risk for an electrical fire.

4. Safety voltage and safety appliances

4.1　Safety voltage

The voltage value without any protective equipment that does not cause harm to the tissues of any part of the human body is called the safety voltage.

Safety voltage regulations in the world include 50 volts, 40 volts, 36 volts, 25 volts and 24 volts, among which 50 volts and 25 volts are the majority. The International Electrotechnical Commission (IEC) limits the safe voltage to 50 volts. In China, the safety voltage level is defined as 12 V, 24 V and 36 V.

Portable lighting used in places with high humidity, narrow, inconvenient movement and large area of ground conductors (such as metal containers, mines, tunnels, etc.) shall be at a safety voltage of 12 volts. In hazardous or especially hazardous environments, portable lighting appliances, local lighting, general lighting should be less than 2.5 meters in height, portable power tools, etc., if no special safety protection device or safety measures, should use safety voltage of 24 V or 36 V.

4.2　Safety appliances

There are three types of safety gear commonly used: insulated gloves, insulated boots, and insulated rods.

1) Insulated gloves

Insulated gloves are made of special rubber with good insulation properties, whether under high voltage or low voltage. They can prevent contacting voltage when operating equipment such as high voltage isolating switch and oil circuit breaker, and working on high voltage appliances and low voltage electrical equipment with live operation.

2) Insulated boots

They are also made of special rubber with good insulation performance, which can prevent the injury of stepping voltage to human body when operating high-voltage or low-voltage electrical equipment.

3) Insulated rod

It is also known as insulating rod, operating rod or brake rod, made of bakelite, plastic, epoxy glass, cloth and other materials.

 New words and phrases

arc extinguishing　灭弧

arrester　*n.*　制动器;避雷器

artificial respiration　人工呼吸

bakelite　*n.*　胶木;人造树胶

cardiopulmonary resuscitation　心肺复苏术

charged body　带电体

chest compression　胸外按压

circuit breaker　断路器

circuit　*n.*　电路;回路

conductive　*adj.*　导电的;传导的

contactor　*n.*　接触器

diffuse　*v.*　扩散;传播

disconnector　*n.*　隔离开关

electric field　电场

electric shock　触电

electrical equipment　电力设备

electrical quantity　电量

electrify　*v.*　使……充电;使……触电

electrocute　*v.*　使……触电受伤;使……触电致死

electrothermal burn　电击烧伤

excitation current　励磁电流;激磁电流

fuse　*n.*　保险丝;熔线

ground resistance　接地电阻

insulate　*v.*　使……绝缘;使……隔热

iron core　铁芯

isolating switch　隔离开关;断路器

live wire　火线

no-load current　空载电流

palpitation　*n.*　心悸;心慌

power outage　停电;动力故障

primary equipment　一次设备;主设备

relay protection　继电保护

respiratory arrest　呼吸停止

secondary equipment　二次设备;辅助设备

static　*adj.*　静态的;静电的

step voltage　跨步电压

step-up   *adj.*   递升的;电压增高的

substation   *n.*   变电站;变电所

syncope   *n.*   晕厥

tachycardia   *n.*   心动过速;心跳过速

tracheal intubation   气管插管

vasoconstriction   *n.*   血管收缩

ventricular fibrillation   心室颤动

 Exercises

Ⅰ. Warming-up questions

Directions：Give brief answers to the following questions.

1. What is the role and scope of electrical primary equipment and secondary equipment?

2. What is the role of high-voltage circuit breakers?

3. What is the purpose of the disconnect switch? How is it classified?

Ⅱ. Technical terms

Directions：Please give the Chinese or English equivalents of the following terms.

1. artificial respiration   2. primary equipment

3. power loss   4. electrical quantity

5. isolating switch   6. charged body

7. 触电   8. 继电保护

9. 绝缘手套   10. 胸外按压

11. 电击烧伤   12. 励磁电流

Ⅲ. Blank filling

Directions：Complete the following sentences by choosing words or phrases given below.

diffuse	static	insulate	step-up	electrify

1. Early electrical workers used it as a coating to _____ coils, and molded it into stand-alone insulators by pressing together layers of shellac-impreg nated paper.

2. When one piece of news on someone's microblog is released, it can _____ like a virus. This tests the government's ability to deal with unexpected incidents.

3. The challenges confronting the drive to _____ automobiles point to a big question underlying the current energy debate in America: how much faith should we have in technology to bail us out?

4. However, many systems are not _____ because their changing behavior cannot be captured by a set of relationships and events.

5. Electromagnetic surge or pulse has blown up _____ transformers in a Russian dam within this past year, and created complete electrical failure in Air France 447.

IV. Translation

Directions: Translate the following passages from English into Chinese or from Chinese into English.

Passage 1

All circuit breaker systems have common features in their operation, but details vary substantially depending on the voltage class, current rating and type of the circuit breaker. The circuit breaker must first detect a fault condition. In small mains and low voltage circuit breakers, this is usually done within the device itself. Typically, the heating or magnetic effects of electric current are employed. Circuit breakers for large currents or high voltages are usually arranged with protective relay pilot devices to sense a fault condition and to operate the opening mechanism. These typically require a separate power source, such as a battery, although some high-voltage circuit breakers are self-contained with current transformers, protective relays, and an internal control power source.

Passage 2

电在造福于人类的同时,也会给人类带来灾难。统计资料表明,在工伤事故中,电气事故占有相当大的比例。以建筑施工死亡人数为例,2011 年全国建筑施工触电死亡人数占其全部事故死亡人数的 7.34%。我国约每用 1.5 亿度电,触电死亡人数 1 人,而美、日等国每用 20 亿—40 亿度电,触电死亡人数约 1 人。据统计,电气火灾约占全部火灾的 20%,造成了巨大的人员伤亡和经济损失。例如,去年北京市发生的 5000 多起火灾中,电气火灾居首位,已成为最大的火灾隐患。

 Additional reading

Unit 8
Smart Grid

With the intensification of global warming and the gradual depletion of fossil energy in recent years, the traditional energy supply system is facing increasing challenges. At the same time, the human society entering the digital network era has also put forward higher requirements on the working performance of traditional power systems, such as transmission efficiency, anti-interference ability, and load capacity. With the rapid development of computer and power electronics technology, smart grid has gradually become a breakthrough direction for the development and innovation of future power systems.

1. The definition of smart grid

What is a smart grid? Smart grid is a concept and vision that captures a range of advanced information, sensing, communications, control, and energy technologies. Taken together, these result in an electric power system that can intelligently integrate the actions of all connected users — from power generators to electricity consumers to those that both produce and consume electricity ("prosumers") — to efficiently deliver sustainable, economic, and secure electricity supplies (see Fig. 8.1).

2. Characteristics of smart grid

Compared with traditional grids, smart grids have strong self-recovery and

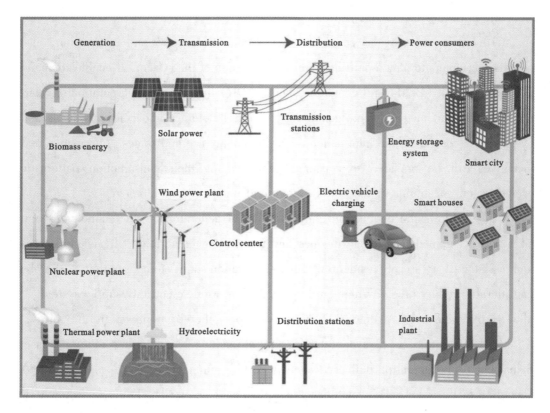

Figure 8. 1   Active participants in the smart grid

Source: https://www.meinbergglobal.com/english/industries/smart-grid-timing.htm.

regulation capabilities. This is because the smart grid can analyze the monitoring data and the collected data fed back by the end user and adjust the configuration of the corresponding parameters in the grid system. In the face of complex natural and social environments, according to the system settings, the smart grid can adjust the system with as little manual intervention as possible to ensure the stability of the power grid and renovate as soon as possible to avoid large-scale paralysis.

Smart grids integrate many forms of clean energy and distributed energy. It can not only adapt to the access of large-scale power generation equipment, but also support the power supply of decentralized power generation equipment. At the same time, it can connect solar energy, wind energy and other clean energy power generation equipment to make the power grid system more green. The cooperation of various technologies guarantees the stability of the power grid.

Smart grid employs innovative products and services along with intelligent control, communication, monitoring and self-healing technologies. The literature suggests the following attributes of the smart grid.

(1) Smart grid provides consumers better choice of supply and information, also permits consumers to play a part in optimizing operation of the system. It enables demand side management (DSM) and demand response (DR) through the incorporation of smart appliances, smart meters, micro-generation, electricity storage and consumer loads and by providing consumers the information regarding energy use and prices. Information and incentives will be provided to consumers for revising their consumption pattern to overcome few constraints in the power system and improving the efficiency.

(2) It allows the connection and operation of generators of all technologies and sizes and accommodates storage devices and intermittent generation. It accommodates and assists all types of residential micro-generation and storage options, thereby considerably reduces the environmental impact of the whole electricity supply system. It allows "plug-and-play" operation of microgenerators, thereby improves the flexibility.

(3) It optimizes and operates assets efficiently by pursuing efficient asset management and operating delivery system (working autonomously, re-routing power) according to the need. This includes the utilizing of assets depending on when it is needed and what is needed.

(4) It operates durably during cyber or physical attacks, disasters and delivers energy to consumers with enhanced levels of security and reliability. It improves and promises security and reliability of supply by predicting and reacting in a self-healing manner.

(5) It provides quality in power supply to house sensitive equipment that enhances with the digital economy.

(6) It opens access to the markets through increased aggregated supply, transmission paths, auxiliary service provisions and DR initiatives.

In general, traditional power grids only conduct power transmission unilaterally, regardless of whether resources are used reasonably or the risk of operation, which inevitably results in waste of resources and large-scale paralysis of the system. The smart grid can integrate the information and resources of each node with the user to realize the real-time monitoring and adjustment of the operation of the entire system. The electricity supplier and the power supplier can communicate without obstacles, and finally realize the entire grid's safety, efficiency, stability and environmental protection.

3. Key technologies of smart grid

Smart grid is an inevitable trend in the development of grid technology. Communication, computer, automation and other technologies have been widely and deeply applied in the power grid and organically integrated with traditional power technologies, which has greatly enhanced the intelligence level of the power grid. To achieve the goals of reliable, safe, economic, efficient, environmentally friendly and safe use of the power grid, some technologies are very important.

3.1   Communication technology

Establishing a high-speed, two-way, real-time, integrated communication system is the basis for realizing a smart grid. Without such a communication system, any smart grid feature cannot be realized, because the data acquisition, protection, and control of smart grids require such a communication system. The establishment of such a communication system is the first step towards a smart grid. At the same time, the communication system must reach as many households as the power grid, thus forming two closely connected networks — the power grid and the communication network. Only in this way can the goals and main features of the smart grid be achieved.

3.2   Measurement technology

Parameter measurement technology is a basic component of the smart grid. Advanced parameter measurement technology obtains data and converts it into data information for use in all aspects of the smart grid. They assess the health of the power grid equipment and the integrity of the power grid, perform meter readings, eliminate electricity cost estimates, prevent power theft, mitigate grid congestion, and communicate with users.

3.3   Equipment technology

Smart grids need to widely apply advanced equipment technology to greatly improve the performance of transmission and distribution systems. The devices in the

future smart grid will fully apply the latest research results in materials, superconductivity, energy storage, power electronics and microelectronics technology, thereby improving power density, power supply reliability and power quality, and efficiency of power production. Through the application and transformation of various advanced equipment, such as equipment based on power electronics technology and new conductor technology, to improve the transmission capacity and reliability of the power grid. Many new energy storage equipments and power sources must be introduced into the power distribution system, while new network structures such as microgrid must be utilized.

3.4　Control technology

Advanced control technology refers to devices and algorithms that analyze, diagnose, and predict states in smart grids and determine and take appropriate measures to eliminate, mitigate, and prevent power interruptions and power quality disturbances. These technologies will provide control methods for transmission, distribution, and user-side control and can manage active and reactive power throughout the power grid. The analysis and diagnosis functions of advanced control technology will introduce a preset expert system, and take automatic control actions within the scope allowed by the expert system. The actions performed in this way will be at the level of one second, and the characteristics of this self-healing grid will greatly improve the reliability of the grid.

3.5　Support technology

Decision support technology transforms complex power system data into understandable information for system operators at a glance. Therefore, animation technology, dynamic coloring technology, virtual reality technology, and other data display technologies are used to help system operators recognize, analyze, and deal with urgent problems. In many cases, the time for system operators to make decisions is reduced from hours to minutes, or even seconds. In this way, smart grids require a broad, seamless, real-time application system, tools, and training to enable grid operators and managers to make decisions quickly.

4. The application of smart grid

By incorporating few technologies, the transition of the conventional electric grid to smart grid is possible. The applications of smart grid technologies are discussed in this section.

4.1　Smart meter

Smart meter is an electricity or gas meter that has metering as well as communication abilities. It measures energy consumption data and permits it to be read remotely and displayed on a device within the home or transmitted securely. The meter can also receive information remotely, e. g. , switch from credit to prepayment mode or to update tariff information (see Fig. 8.2). It has two key functions to perform: (1) providing data on energy usage to consumers to help control over consumption and cost; (2) sending data to the utility for peak-load requirements, load factor control and to develop pricing strategies on the basis of consumption information. Key features of smart meters are automated data reading and two-way communication between utilities and consumers. Smart meters are developed to measure electricity, gas and water consumption data. In smart grid, smart meters provide consumers with knowledge about how and when they use energy and how much they pay for per kilowatt hour of energy.

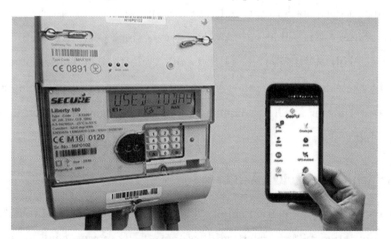

Figure 8.2　The role of smart meters in utilities

Source: en. wikipedia. org.

This will result in better pricing information and more accurate bills and it will guarantee faster outage detection and restoration by the utility. Additional features of smart meters include tariff options, tax credits, DR rates, smart thermostat, prepaid metering, switching, enhanced grid monitoring, remote connect / disconnect of users, appliance control and monitoring and participation in voluntary rewards programs for reduced consumption. Smart meter outputs can be used for voltage stability and security assessment.

4.2　Automated meter reading

Automated meter reading (AMR) devices let utilities to read meters remotely, removing the requirement to send a worker to read each meter separately. While they do represent a certain amount of two-way communication, this functionality is limited and does not increase the efficiency or reliability of the utility grid. They do not have any inbuilt home displays to show the energy consumption pattern to the consumer, hence consumer remains unaware of their energy consumption. As a result, utility companies cannot notify consumers of their energy consumption, thus consumers cannot replenish their energy consumption during peak hours and save energy. AMR in the distribution network lets utility companies read the status from consumers' premises, alarms and consumption records remotely. Capability of AMR is restricted to reading meter data due to its one-way communication system. Based on the information received from the meters it does not let utilities take corrective action. In other words, transition to the smart grid is not possible with AMR systems, since pervasive control at all levels is not possible with AMR alone. AMR is the collection of consumption data from consumers like electric meters and smart meters remotely using telephony, radio frequency, powerline or satellite communications technologies and process the data to generate the bill.

4.3　Vehicle to grid (V2G)

The incorporation of electric vehicles and plug-in hybrid electric vehicle (PHEV) is another application of the smart grid system. V2G power employs electric-drive vehicles to provide power to particular electric markets. Fuel cells, battery or hybrid of these two is employed to store energy in vehicles. There are three main different versions of the V2G concept.

(1) A hybrid or fuel cell vehicle.

(2) A battery-powered or plug-in hybrid vehicle.

(3) A solar vehicle, of which involves an onboard battery.

The major advantages of V2G are as follows. It provides storage space for renewable energy generation and it stabilizes large scale wind generation via regulation. PHEV significantly cuts down the local air pollution problems. Hybridization of electric vehicles and associations to the utility grid conquers the limitations of their use including battery weight / size, cost and short range of application. PHEV offers an alternative to substitute the use of petroleum based energy sources and to reduce overall emissions by using a mix of energy resources. The use of PHEVs potentially has a significant positive impact on the electric power system from the point of view of increasing electric energy consumption, substituting petroleum fuels with unconventional sources of energy. The associations between vehicles and the utility grid are illustrated in the Fig. 8. 3.

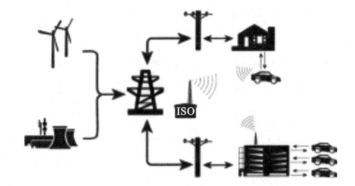

Figure 8. 3   The connections between vehicles and the utility grid

Source: innopaths. eu.

4. 4   Plug-in hybrid electric vehicle technology

A PHEV is a hybrid electric vehicle with a larger battery pack. So, it runs on electricity when its battery SOC is high or else, the IC engine takes over and the vehicle uses gasoline similar to a hybrid vehicle. The battery pack can be recharged via a plug which provides connection to the utility grid; hence, a PHEV, compared to conventional cars, has an extra equipment to connect to an external electrical source for recharging. PHEVs are characterized by their all-electric range. In cases of extreme emergencies like a sudden increase in oil prices or major decrease in oil supplies, the stored or unused energy that utilities preserve during night time or off-peak time can be utilized to support the vehicles. It must be considered that efficiency of electric drive systems is about 70% only, as an example, a first-generation PHEV can travel about 75 cents per gallon of gas or about 3-4 miles per kWh. All PHEV will be employed with connection to the utility grid for electrical energy flow, a logical connection or control is compulsory for communication with the utility grid operator and onboard metering and controls. Fuel cells can generate power from gaseous and liquid fuels and PHEV can function in either capacity.

PHEV operates in three modes: charge-sustaining, charge-depleting and blended mode or mixed modes. These vehicles can be designed to drive for an extended range in all-electric mode, either at low speeds only or at all speeds. These modes manage the vehicle's battery discharge strategy and their use has a direct effect on the size and type of battery required. In charge-sustaining mode certain amount of charge above battery SOC is sustained for emergency use. Before reaching SOC, vehicle's IC engine or fuel cell will be engaged. Charge-depleting mode permits a fully charged PHEV to operate exclusively on electric power until its battery SOC is depleted to a predetermined level, at which time the vehicle's IC engine or fuel cell will be engaged. This period is the vehicle's all-electric range. This is the only mode that a battery electric vehicle can operate in, hence their limited range. Mixed mode describes a trip using a combination of multiple modes. For example, a car may begin a trip in low speed charge-depleting mode, then enter onto a freeway and operate in blended mode. The driver might exit the freeway and drive without the IC engine until all-electric range is exhausted. The vehicle can revert to a charge sustaining-mode until the final destination is reached. This contrasts with a charge-depleting trip which would be driven within the limits of a PHEV's all-electric range.

Advantages of PHEV are as following.

(1) Operating costs.

(2) Vehicle-to-grid electricity.

(3) Fuel efficiency and petroleum displacement.

Disadvantages of PHEV are as following.

(1) Cost of batteries.

(2) Recharging outside home garages.

（3）Emissions shifted to electric plants.

（4）Tiered rate structure for electric bills.

（5）Lithium availability and supply security.

4.5　Smart sensors

Smart sensors are defined as sensors that provide analog signal processing of recorded signals, digital representation of the analog signal, address and data transfer through a bidirectional digital bus, manipulating, and computing of the sensor-derived data. A smart sensor provides additional functions further than those required for generating an accurate demonstration of the sensed quantity. It is composed of many processing components integrated with the sensor on the same chip and has intelligence of some forms and provides value-added functions beyond passing raw signals, leveraging communications technology for telemetry and remote operation / reporting. Objectives of smart sensors consist of integrating and sustaining the distributed sensor system measuring intelligently and smartly, crafting a general platform for controlling, computing, yielding cost effectiveness and communication toward a common goal and interfacing different type's sensors. The virtual sensor is a component of the smart sensor, which is a physical transducer / sensor, plus a connected digital signal processing (DSP) and signal conditioning necessary for obtaining reliable estimates of the essential sensory information.

Smart sensors enable more accurate and automated collection of environmental data with less erroneous noise amongst the accurately recorded information. It offers functionalities beyond conventional sensors through fusion of embedded intelligence to process raw data into actionable information that can trigger corrective or predictive actions. Smart sensors are extensively employed in monitoring and control mechanisms in variety of fields including smart grid, battle field, exploration and a great number of science applications. For supporting smart grid monitoring and diagnostics applications, automated, reliable, online and off-line analysis systems are required in conjunction with smart sensors. Smart sensors enable condition monitoring and diagnosis of main substation and line equipment including transformers, circuit breakers, relays, cables, capacitors, switches and bushings. Figure 8.4 shows the basic block diagram of smart sensor.

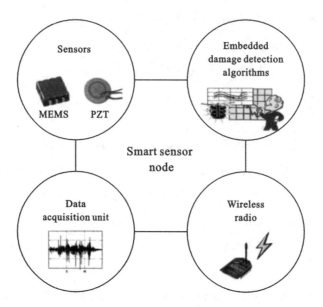

Figure 8. 4　Components of smart sensor node

Source: https://www. researchgate. net/publication/263400026 _ Development _ of _ Acceleration-PZT _ Impedance _ Hybrid_Sensor_Nodes_Embedding_Damage_Identification_Algorithm_for_PSC_Girders/figures? lo = 1&utm_source = google&utm_medium = organic.

5. The future: the key challenges of smart grid

The major challenges that smart grid facing are as follows.

(1) Strengthening the utility grid. It must be ensured that the utility grid has sufficient transmission capacity to accommodate more energy resources, especially renewable resources.

(2) Moving offshore. Most effective and efficient connections for offshore wind farms and for other marine technologies (tidal and wave energy) which are stochastic in nature, must be developed.

(3) Developing decentralized architectures. Decentralized architectures must be developed to enable harmonious operation of small-scale electricity supply systems with the total system.

(4) Communications. Developing a communication infrastructure which allows the operation and trade of potentially millions parties in a single market.

(5) Active demand side. Enabling all consumers to play an active role in the operation of the system, with or without their own generation.

（6）Integrating intermittent generation. Finding the best ways for integrating intermittent generation like residential micro-generation.

（7）Enhanced intelligence of generations. The problems associated with enhanced intelligence generation schemes （like FREEDM） system must be resolved to revolutionize the utility grid.

（8）Advanced power system monitoring, protection and control. Advanced measurement schemes like synchronized phasor measurements must be common to achieve synchronization by the same time.

（9）Capturing the benefits of DG and storage. Advanced technologies must be developed to capture DERs more effectively. Hybrid energy system, such as SPV-wind, SPV-fuel cells are necessary to maintain reliability and to power remote areas.

（10）Preparing for electrical vehicles. Electrical vehicles are mostly emphasized due to their mobile and highly dispersed character and possible massive employment in the next years, which would yield a key challenge.

 New words and phrases

aggregate　　*v.*　　集合;聚集

algorithm　　*n.*　　算法;运算法则

attribute　　*n.*　　属性;特质

battery pack　　电池组

battery-powered　　*adj.*　　靠电池供电的

bidirectional　　*adj.*　　双向的;双向作用的

capture　　*n.*　　俘虏;捕获

charge-depleting　　*n.*　　电量消耗

charge-sustaining　　*n.*　　电量保持

conductor　　*n.*　　导体

configuration　　*n.*　　配置;结构

data acquisition　　数据采集

decentralized　　*adj.*　　分散管理的

depletion　　*n.*　　消耗;损耗

disperse　*v.*　分散;使……散开

electric meter　电表

fuel cell　燃料电池

hybridization　*n.*　杂交;配种

incentive　*n.*　动机;刺激

load capacity　负载能力

load factor　负荷系数;载荷因素

micro-grid　*n.*　微电网

mitigate　*v.*　使……缓和;使……减轻

onboard battery　车载电池

optimize　*v.*　使……最优化;使……完善

outage　*n.*　（电力等）停止供应期;断供期

parameter measurement　参数测量

peak-load　*n.*　最大负荷;峰荷

plug-in hybrid electric vehicle　插电式混合动力汽车

power density　功率密度

power line　电源线;输电线

power quality　电能质量

power source　电源;能源

seamless　*adj.*　无缝的;无缝合线的

smart meter　智能电表

superconductivity　*n.*　超导电性

thermostat　*n.*　恒温器;自动调温器

transmission capacity　输电量

unilaterally　*adv.*　单方面地

utility　*n.*　设施;公共事业

 Exercises

Ⅰ. Warming-up questions

Directions：Give brief answers to the following questions.

1. What makes smart grid "smart"?

2. What technologies enable smart grid to operate efficiently?

3. How can smart grid be applied in different fields? Please give an example.

Ⅱ. Technical terms

Directions: Please give the Chinese or English equivalents of the following terms.

1. peak-load
2. plug-in hybrid electric vehicle

3. superconductivity
4. data acquisition

5. load capacity
6. parameter measurement

7. 功率密度
8. 输电量

9. 燃料电池
10. 车载电池

11. 电表
12. 能源损耗

Ⅲ. Blank filling

Directions: Complete the following sentences by choosing words or phrases given below.

disperse　　aggregate　　seamless　　bidirectional　　depletion

1. Consumers have to expend extra effort to _____ these too finely grained services to realize any benefit as well as have the knowledge of how to use these services together.

2. If your information requires a more _____ flow of information across topic boundaries, don't use this architecture.

3. These actions indicate the multiplicity of the source and target types as well as the navigability from the source object, either _____ or unidirectional.

4. The bank's technical assistance will also promote the use of life cycle costing —— a method that looks at total costs, including resource _____ and environmental impact.

5. The pollutants _____ easily across wide geographic areas, retain their toxicity, and have a tendency to accumulate in the fatty tissues of organisms.

Ⅳ. Translation

Directions：Translate the following passages from English into Chinese or from Chinese into English.

Passage 1

In renewable energy, smart grid is a sector or a communication area that can connect the production from renewable energy sources to the grid. However, the communication between renewable energy production and smart grid brings many challenges such as stability issues, complicated operating process and remote control together. The grid is not only electrical transmission system from power plants to the substation, but it also covers the distribution, electricity from the substations to the individual user. There will be many challenging processes and technologies included in the smart grid system, such as monitoring and analysis, automation or control (active control of high voltage device, robustness, reliability, security and efficiency, etc.), integration and control of distributed energy resources such as micro grid, renewable energies, solid oxide, fuel cells, battery storage systems, and etc.

Passage 2

智能电网的研发几乎涵盖了目前所有正蓬勃发展的新兴产业,包括计算机、新能源、电子信息、通信技术、系统控制技术等高新技术领域。可以预见的是,未来智能电网有可能像之前兴起的互联网一样,又一次对各个行业产生不可估量的深远影响,甚至可能引发能源系统新一轮产业转型。未来智能电网将在无人化、信息化的方向不断进行技术突破和结构革新,智能电网所覆盖的范围也将逐渐扩大。

 Additional reading

Unit 9

Energy and Policy

Since the late 1970s China has made remarkable progress in energy industry. China has become the world's largest energy producer, with a well developed energy supply system for coal, electricity, oil and gas, as well as new and renewable sources of energy. Service for energy consumption has been greatly improved. Meanwhile, a series of hardships harass us: shortage of energy, low energy efficiency, environmental pollution, energy security, reform in mechanism of energy system, and etc. All these challenge our policy for the development of energy.

According to the Chinese Energy Policy issued by the Information Office of the State Council in 2012 (see Fig. 9.1), the basic content of China's energy policy is to adhere to the energy development policy of "saving first, basing ourselves on domestic conditions, diversified development, environmental protection, scientific and technological innovation, deepening reform, international cooperation and improving people's livelihood".

Figure 9.1   The White Paper of Chinese Energy Policy (2012)

Source: https://baike.so.com/doc/
6836977-7054215.html.

1. Policy about energy conservation

China has a large population and relatively insufficient resources. In order to realize the sustainable utilization of energy resources and the sustainable development of economy and society, we must take the road of energy conservation. Energy conservation has always been a top priority in China.

1.1　Optimizing the industrial structure

China adheres to the adjustment of industrial structure as the strategic focus of energy conservation. We will strictly control low-level redundant construction and accelerate the elimination of backward production capacity with high energy consumption and emissions, speed up the application of advanced applicable technology to upgrade traditional industries, and raise the threshold of access to processing trade and promote the transformation and upgrading of processing trade. We will improve the structure of foreign trade and promote its development from being energy-and-labor-intensive to being capital-and-technology-intensive. We will promote the development of the service sector. We will foster and develop strategic emerging industries and accelerate the formation of leading and pillar industries.

1.2　Strengthening energy conservation in industry

Industrial energy consumption accounts for more than 70 percent of China's energy consumption, and industry is a key area for energy conservation. The state formulates a catalogue of advanced and applicable technologies for energy conservation and emission reduction in key industries such as iron and steel, petrochemical, non-ferrous metals and building materials, eliminates outdated processes, equipment and products, and develops energy-saving and high value-added products and equipment. We will establish and improve a mandatory system of standards for energy consumption quotas for products of key industries and strengthen an energy-saving assessment and review system. We will launch and implement key energy-saving projects, such as cogeneration of heat and power, recovery and utilization of coal gas from industrial by-products, construction of energy management and control centers of enterprises, and cultivation of energy-saving industries, so as to raise the energy efficiency of enterprises.

1.3   Improvement of building energy efficiency

The country vigorously develops the green building, establishing and perfecting Green Building Standard, carrying out green building rating and marking. More efforts will be made to promote energy-saving renovation of existing buildings, implement a public notice system for energy consumption and efficiency of public buildings, establish a life-cycle management system for the use of buildings, and strictly manage the demolition of buildings. We will formulate and implement energy-saving plans for public institutions and strengthen the building of an energy-saving regulatory system for public buildings. We will carry forward the heat metering and energy-saving renovation of existing buildings in northern heating areas, implement the "energy-saving greenhouse" (see Fig. 9.2) project, rebuild the old heating network, and implement the management of heat metering charges and energy consumption quota.

Figure 9.2   An idea of how to promote building energy efficiency

Source: http://www.yclunwen.com/gongchengjiegou/96015.html.

1.4 Promoting energy-saving in transportation

Priority will be given to public transport. We will actively promote the construction of intercity rail transit, and reasonably advocate green travel; implement world-class fuel consumption standards for automobiles and promote the use of energy-saving and environmental-friendly means of transport; accelerate the elimination of old cars, locomotives, ships; optimize the structure of transportation, and vigorously develop green logistics, increase the proportion of railway electrification and carry out energy-saving reconstruction of airports, wharves and stations. We will actively promote research, development and application of new energy vehicles, and make scientific plans and construction of supporting facilities such as gas filling and charging facilities (see Fig. 9.3).

Figure 9.3 A model of new energy vehicle at an auto show

Source: https://www.chinadaily.com.cn/bizchina/2017-02/06/content_28111037.htm.

1.5 Motivating energy conservation nationwide

We will step up energy-saving education and publicity, encourage and guide urban and rural residents to form green consumption patterns and lifestyles, and raise the awareness of the whole nation about energy conservation. We will strictly enforce energy-saving standards and norms for public institutions and give full play to the exemplary role of government agencies; mobilize broad participation of all sectors of society, actively carry out energy-saving actions in residential areas, schools, government organs, military barracks and enterprises, and strive to establish a long-term mechanism of energy conservation for the whole society. We will promote energy-saving and emission reduction in agriculture and rural areas and promote the construction of energy-saving residential buildings.

2. Policy about new and renewable energy sources

Vigorously developing new and renewable energy sources is an important strategic measure to promote clean and diversified energy development and cultivate strategic emerging industries. It is also an urgent need to protect the ecological environment, cope with climate change and achieve sustainable development. China is firmly committed to developing new and renewable energy sources. By the end of the 12th five-year plan, non-fossil energy consumption will account for 11.4 percent of primary energy consumption, and the proportion of installed non-fossil energy power generation will reach 30 percent.

2.1 Hydroelectric power

As demonstrated in Figure 9.4 and Figure 9.5, world hydroelectric power generation has risen steadily by an average 3 percent annually over the past four decades. Four countries dominate the hydropower landscape: China, Brazil, Canada, and the United States. Together they produce more than half of the world's hydroelectricity.

China is rich in hydropower resources, with a technical exploitable capacity of 542 million kilowatts, ranking first in the world. In terms of electricity generation, hydropower generation accounts for less than 30%, and still has great potential. To meet the 2020 target of 15 percent share of non-fossil energy consumption, more than half will be dependent on hydropower. For the sake of ecological and environmental protection and resettlement, China will actively develop hydropower, combining hydropower development with the promotion of local employment and economic development. We will improve policies on resettlement for hydropower immigrants and

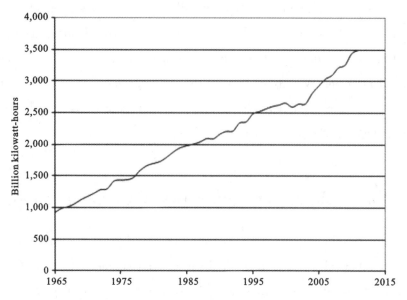

Figure 9. 4　World hydroelectric generation：1965-2011

Source：https：//www. treehugger. com/renewable-energy/hydropower-continues-steady-growth. html.

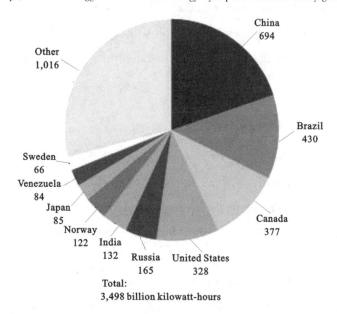

Figure 9. 5　Hydroelectric generation by country, 2011（billion kilowatt-hours）

Source：https：//www. treehugger. com/renewable-energy/hydropower-continues-steady-growth. html.

improve the mechanism for sharing benefits. We will strengthen the protection of the ecological environment and the environmental impact assessment, strictly implement the ecological protection measures for the built hydropower stations, and improve the comprehensive utilization of water resources and the ecological and environmental benefits. It is necessary to do a good job in hydropower development basin planning, speed up the construction of large-scale hydropower stations in key river basins, develop hydropower resources in small and medium-sized rivers according to local conditions, and make scientific planning and construction of pumped-storage hydroelectricity.

2.2 Nuclear power

As a kind of clean, efficient and high-quality modern energy, nuclear power plays an important role in optimizing the energy structure and ensuring the national energy security. In the 2000s, China produced only 1.8% of its electricity from nuclear power, well below the world average of 14%. Nuclear safety is the lifeline of nuclear power development. After the Fukushima nuclear accident, China launched a comprehensive and rigorous safety inspection of its nuclear power plants. The inspection results show that China's nuclear power safety is guaranteed. We will continue to implement the principle of "safety first" in the whole process of nuclear power planning, site selection, R&D, design, construction, operation and decommissioning.

To formulate and improve the nuclear power regulatory system, we will improve and optimize the safety management mechanism for nuclear power, strictly set the threshold for access, and fulfill the responsibilities of safety entities, strengthen safety supervision and inspection of nuclear power plants under construction and operation, and strengthen environmental supervision and management of radiation. A national nuclear accident emergency mechanism shall be established and improved to enhance emergency response capabilities. We will increase input in nuclear power science and technology innovation, promote the application of advanced nuclear power technology, improve nuclear power equipment, and pay attention to the training of nuclear power talents.

2.3 Wind power

At present, wind power is the best choice of non-hydro renewable energy for large-scale development and market-oriented utilization. China is the world's fastest-growing wind power country. In terms of optimizing the distribution of wind power exploitation,

we have promoted wind power construction in areas rich in wind energy resources in northwest, north and northeast China in an orderly manner and accelerate the development and utilization of decentralized wind energy resources. We will boost steady development of offshore wind power. Improve wind power equipment standards and industrial monitoring system. We will also encourage wind power equipment enterprises to strengthen research and development of key technologies and accelerate the technological upgrading of the wind power industry, enhance power grid capacity of wind energy dissipation by strengthening the construction of power grid, improving power grid dispatching and the performance of wind power equipment, strengthening wind power forecasting, and etc.

2.4 Solar energy

Abundant with solar energy resources, China has great potential and broad application prospects in development of solar energy. Large-scale grid-connected photovoltaic power stations and solar thermal power projects are being built in areas with abundant solar energy resources, such as Qinghai, Xinjiang, Gansu and Inner Mongolia. The construction of building-integrated distributed photovoltaic power generation systems in the central and eastern regions are encouraged. We will increase the popularity of solar water heating and encourage the use of solar energy for central water heating, solar heating and cooling, and applications of solar energy for medium and high temperature industries; promote the use of solar water heating, solar cookers and solar houses in rural areas, border areas and small towns.

2.5 Other new energy sources

China adheres to the principle of "taking into account all factors, adapting measures to local conditions, comprehensive utilization and orderly development", and develops biomass energy and other renewable energy sources. In the main grain and cotton yielding areas, we develop biomass power generation, making full use of the crop straw, grain processing residues and bagasse as fuel. We will explore forest biomass power generation in areas rich in forest resources, develop municipal solid waste incineration and landfill gas power generation, promote biogas and other biomass gas supply projects in areas where conditions permit, build biomass briquettes production base according to local conditions. On the premise of protecting groundwater resources, the technology of high-efficiency utilization of geothermal energy is popularized. We will strengthen the follow-up and research on the development and utilization of tidal energy, wave energy and hot dry rock power generation.

3. Policy about fossil energy

From a global perspective, coal, oil and other fossil fuels will remain the main source of energy supply for a long time, and China is no exception. China will make overall plans for the development and utilization of fossil energy and environmental protection, accelerate the building of advanced production capacity, eliminate backward production capacity, vigorously promote the clean development of fossil energy, protect the ecological environment, respond to climate change, and achieve energy conservation and emission reduction.

3.1　Coal

China's coal industry adheres to the principle of safety and efficiency exploitation and utilization of source, not to neglect environmental protection. Priority is given to the construction of large and modern open-pit coal mines and super-large mines. The upgrading and transformation of coal mines and the elimination of backward production capacity have been carried out, and the degree of coal mining mechanization and the level of safe production have been raised. We will vigorously develop the circular economy of mining areas, increase the proportion of coal washing and dressing, and rationally develop the associated resources of coal. In accordance with the development orientation of energy-intensive, technology-intensive, capital-intensive, long industrial chain and high added value, the demonstration project of coal deep processing and upgrading shall be built in an orderly manner. We will encourage the construction of projects for the clean utilization and processing transformation of low calorific value coal, strengthen environmental protection and ecological construction in coal mining areas, and do a good job in comprehensive ecological management and land reclamation in coal mining subsidence and affected areas.

3.2　Thermal power

China adheres to the principle of low carbon, clean and high efficiency and vigorously develops green thermal power. We will encourage the integrated development of coal and electricity, and steadily promote the construction of large coal and electricity bases. We will actively apply advanced power generation technologies such as supercritical and ultra supercritical technologies to build clean and efficient coal-fired units and energy-saving and environmental-friendly power plants. We will continue to phase out energy-consuming and polluting small thermal power units, strictly control pollutant emissions from coal-fired power plants. Newly-built coal-fired power units shall be installed with facilities for dust removal, desulphurization and denitration simultaneously. We will encourage the construction of cogeneration units in areas with concentrated heat loads, such as large and medium-sized cities and industrial parks, construct gas-steam combined cycle peaking units in suitable areas, and actively promote the combined supply of natural gas, heat, power and cooling. The popularization and application of water-saving technology should strengthened in thermal power plants, and projects concerned Integrated coal Gasification Combined Cycle (IGCC), as well as carbon capture and storage technology.

3.3　Conventional oil and gas resources

China will continue to pursue the policy of simultaneous development of oil and gas, stabilize the east, speed up the west, develop the south and explore sea areas. We will steadily increase crude oil reserves and stabilize oil production, and steadily promote the exploration and development of key oil production areas, such as Tarim Basin, Ordos Basin, strengthen the reformation in old oil fields to enhance oil recovery. We will accelerate the development of natural gas, increase the production capacity of the midwestern main gas fields, promote the exploration and development of offshore oil and gas fields, and gradually increase the proportion of natural gas in the primary energy structure, optimize the layout of oil refining industry, building several large-scale oil refining and chemical bases, form the big oil refining agglomeration areas around Bohai Sea, Yangtze River Delta and Pearl River Delta, realizing the integration of upstream and downstream, oil refining and chemical industry, and oil refining and storage.

3.4   Unconventional oil and gas resources

Unconventional oil and gas resources play an indispensable role in China's energy supply. China will accelerate the exploration and development of coal-bed methane, increase proved geological reserves and promote the construction of industrial bases of coal-bed methane such as the Qinshui Basin and the eastern margin of the Ordos Basin. We will speed up the exploration and development of shale gas, select a number of shale gas prospect areas and favorable target areas. We will accelerate the capture of core technologies for shale gas exploration and development, establish new mechanisms for shale gas exploration and development, implement industrial incentive policies and improve supporting infrastructure, so as to lay a solid foundation for the rapid development of shale gas in the future. We will intensify the exploration and development of such unconventional oil and gas resources as shale oil and oil sands.

3.5   Construction of energy storage and transportation facilities

We will make overall plans for the construction of energy transmission channels by taking into account such factors as the target market, the adjustment of the industrial distribution, as well as the water and ecological carrying capacity of the resource areas, and etc. We will speed up the expansion and renovation of existing trunk railway lines and the construction of new railway coal transport corridors, improve coal transport capacity across regions, and build supporting ports and docks. We will further expand the transmission of electricity from the west to the east and from the north to the south, improve regional main power grids, develop advanced transmission technologies such as ultra-high voltage, and improve the capacity of optimal allocation of power grid resources. We will strengthen the construction of main pipelines for crude oil, refined oil and natural gas, improve the regional transportation network, and build large-scale oil and gas terminal stations along the coast, strictly implement the laws and regulations on the protection of oil and gas pipelines to ensure the safe operation of oil and gas pipelines. We will pool state and commercial reserves, strengthen capacity-building for emergency support, and improve the reserve systems for crude oil, refined oil, natural gas and coal, improve peak-shaving capacity of natural gas, establish and perfect coal peak-shaving reserve.

4. Energy management system

Reform is a powerful driving force for accelerating the transformation of the development mode. China will firmly advance the reform in the field of energy, strengthen top-level design and overall planning, accelerate the establishment of an institutional mechanism conducive to the scientific development of energy, and push forward the transformation of the mode of energy production and utilization, to safeguard the nation's energy security.

4.1 Legal system for energy

Despite hardships to overcome in energy legislation, China attaches great importance to and continues to actively promote the development of the energy legal system. In the 21st century, China is studying and demonstrating the enactment of the energy law, as well as administrative regulations on oil reserves, protection of offshore oil and gas pipelines, and management of nuclear power. We will revise and improve existing laws and regulations, including the coal laws and the electric power laws, and advance legislation in the fields of oil, natural gas and atomic energy.

4.2 Market system and mechanism

China is actively promoting energy market-oriented reform and giving full play to the fundamental role of the market in resource allocation. All projects included in the national energy plan are open to private capital except those explicitly prohibited by laws and regulations. We will encourage private capital to participate in the exploration and development of energy and resources, the construction of oil and gas pipeline networks and power construction, and encourage private capital to develop the coal processing and conversion and oil refining industries. We will continue to support private capital to fully enter the new and renewable energy industries. We will strengthen and standardize the management of coal exploration and development rights, gradually abolish the dual-track system of prices between key contract coal and market coal, and improve the coordinated development mechanism between coal and coal-bed methane. We will deepen reform of the power system and steadily carry out pilot projects to separate transmission and distribution. We will actively carry forward the reform of electricity tariffs and gradually establish a price mechanism whereby the prices of electricity generation and sale are set by the market and the prices of transmission and distribution are set by the government, straighten out relationship between the price of coal and electricity. We successfully carried out the reform of the linkage of oil prices, taxes and fees, and guided energy consumption reasonably by means of taxation. We will continue to improve and straighten out the pricing mechanism for refined oil products and carry out pilot reforms of the pricing mechanism for natural gas. We will improve the energy market system and develop spot, long-term contracts, futures and other trading forms.

4.3   Management of the energy sector

To improve the efficiency of energy resources development and utilization, promote the scientific development of energy industry and safeguard national energy security, we must strengthen the management of energy industry. We should attach importance to the strategic planning and macroeconomic regulation and control of energy development and implement industry management by means of integrated planning, policies and standards. By reducing the government's intervention in micro-affairs and simplifying the administrative examination and approval matters, we will strengthen regulation of monopoly and unfair competition practices and establish an open, fair, scientific and effective regulatory system. We will strengthen the prediction and management of energy statistics and improve the energy statistics, monitoring, prediction and early warning system.

Abiding by the directive policies and various adaptable regulations to safely and efficiently explore and utilize energy resources, and seeking for worldwide cooperation, China strive to forge ahead against any challenges and stick to sustainable development of energy.

 New words and phrases

agglomeration   *n.*   凝聚;结块

charging post　充电桩

coal-bed methane　煤层气

demolition　*n.*　拆除;破坏;毁坏

denitration　*n.*　脱硝;脱硝酸盐作用

enactment　*n.*　制定;颁布;通过

energy-saving renovation　节能改造

exploitable　*adj.*　可开发的;可利用的

forge ahead　继续进行;取得进展

formulate　*v.*　制订;规划;构想

harass　*v.*　骚扰;使……困扰（或烦恼）

high-end equipment　高端设备

integrated gasification combined cycle　整体煤气化循环发电

low calorific value　低卡值;低热值

non-ferrous metal　有色金属

offshore wind power　海上风电

oil refining　炼油

open-pit　*adj.*　露天开采的

phase out　逐步淘汰;逐渐停止

premise　*n.*　前提;假定

resource allocation　资源配置

shale gas　页岩气

strategic emerging industry　战略新兴产业

ultra-supercritical　*adj.*　超临界

 Exercises

Ⅰ. Warming-up questions

Directions：Give brief answers to the following questions.

1. Can you list some challenges that we have encountered in energy development? The more specific, the better.

2. What roles can the policy-makers, corporations and individuals play in facing the above challenges?

3. Why is it necessary to promote international cooperation in energy deployment?

Ⅱ. Technical terms

Directions: Please give the Chinese or English equivalents of the following terms.

1. energy efficiency   2. oil recovery

3. renewable energy quota   4. phase out

5. electric vehicle   6. offshore wind power

7. 节能减排   8. 环境友好型社会

9. 淘汰落后产能   10. 节能改造

11. 高端设备   12. 战略新兴产业

Ⅲ. Blank filling

Directions: Complete the following sentences by choosing words or phrases given below.

curb	incompatible with	phase out	availability	concerted

1. There is a limit to what any one company can do when it operates within a system in which oil, gas, and even coal use are still rising globally — even as the _____ of renewable energy sources grows dramatically.

2. Although stronger climate action is most welcome, to truly make a dent in global carbon emissions, both corporations and activists need to shift their focus to engage a much broader set of politicians and businesspeople who have the most potential to _____ emissions.

3. In particular, there needs to be greater focus on _____ coal in emerging markets. Coal remains responsible for around 30 percent of global energy-related carbon dioxide emissions — and its use is projected to remain stable.

4. It will require _____ work by the companies, greater transparency about the results of their actions, and a cooperative spirit in civil society.

5. Despite the world's growing need for affordable energy, continued growth in emissions from hydrocarbons is _____ climate goals. At current emission rates, the carbon budget to limit warming to a 1.5 degree Celsius target will be used up in around 15 years.

Ⅳ. Translation

Directions: Translate the following passages from English into Chinese or from Chinese into English.

Passage 1

The dramatic increase in urgency and attention to the issue of climate change last week at Davos was very welcome, but turning ambition into action will require that corporate and activist leaders alike increasingly advocate for policy change, focus on the regions and energy sectors driving emissions growth, and find common ground on the best role for the energy industry to play in delivering decarbonization solutions.

Passage 2

加强能源出口国、消费国和中转国之间的对话和交流,是开展能源国际合作的基础。国际社会应进一步密切双、多边关系,加强在提高能效、节能环保、能源管理、能源政策等方面的对话交流,完善国际能源市场监测和应急机制,深化在信息交流、人员培训、协调行动等方面的合作。

 Additional reading

Unit 10
Energy and Technology

Technological innovation has the potential to disrupt energy markets and influence energy policy around the world. It also plays an important role to improve access to energy sources in developing nations and to meet global environment goals. Energy science and technology focus on energy efficiency, clean energy and energy environment technology.

In June 2014, Comrade Xi Jinping made an important deployment on China's energy security strategy, putting forward the "four revolutions" in the energy sector, including the revolution in energy consumption, the revolution in energy supply, the revolution in energy technology and the revolution in energy system. China's initiative to explore the construction of a global energy internet is a comprehensive innovation in the field of energy from three aspects: energy transmission, energy dispatching and energy supply.

The global energy internet is a global energy configuration platform based on ultra-high voltage (UHV), ubiquitous smart grid and clean energy transmission. UHV power grids can transmit large amounts of power over long distances with low losses, thus achieving more economical trans-regional and trans-continental transmission, and a reliable, safe, economic, efficient and environmentally friendly power grid. Clean energy includes water energy, wind energy, solar energy, nuclear energy, marine energy, bio-energy, and etc., in theory, only a small part of which could be exploited to meet all of humanity's current needs. With ultra-high voltage power grid as a means of energy transmission, ubiquitous smart grid as a major dispatching platform and clean

energy as a core supply resource, the global energy internet will become an interconnected energy network.

The development of the global energy network and the trend of technological innovation is beneficial to the common response of mankind to the increasingly severe challenges of resources and environment.

1. Energy efficiency and energy-saving technology

With the rapid economic growth and active human activities, global demand for energy has been drastically increasing and it is crucial to make good use of the limited resources and exploit new sources of reliable affordable energy.

1.1 Enhanced oil recovery (EOR)

At present, the world average oil recovery rate is 35%, enhanced oil recovery technology can increase the average recovery rate to 50% up to 65%-80% . This means that the world's developed oilfields could acquire 120bn more tonnes reserves and 40bn tonnes of potential reserves.

In Daqing Field, China, such innovative tertiary recovery technologies as asp flooding and polymer flooding have resulted in a recovery rate of over 50 percent in the main fields, 15 percentage points higher than that of similar oilfields at home and abroad. Adopting polymer flooding technology expect to increase 300 Mt of reserves impossible to extract with conventional techniques. Figure 10. 1 shows the EOR numerical simulation of gas-water two-phase mixing injection.

1.2 Supercritical and ultra-supercritical thermal power units

The improvement and application of supercritical and ultra-supercritical thermal power generation technology will greatly improve power generation efficiency and reduce pollutant emissions. Thermal efficiency of Denmark's coal-fired ultra-supercritical units reaches as high as 47%. US coal-fired ultra-supercritical unit maximum capacity has reached 1300 MW. In March 2007, China's first home-made 1000 MW ultra-supercritical unit was put into commercial operation, with its coal consumption 77. 5 g/kWh less than the national average coal consumption in 2006.

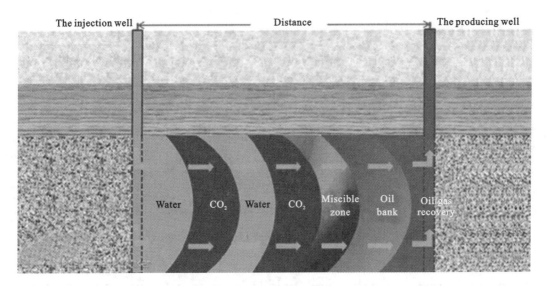

Figure 10. 1   EOR numerical simulation of gas-water two-phase

mixing injection

Source: https://www. sohu. com/a/54402905_232491.

The research and development of ultra-supercritical units are mainly focused on high-temperature resistant alloy steels. The ferritic heat-resistant steels are being developed in Japan. The United States is developing high-temperature high-strength alloy steel at 760 ℃ for the coal-fired ultra-supercritical units under the project of Vision 21.

1. 3   Ultra-high voltage transmission

According to China's grid voltage standards, AC rated voltage over 1000 kV, or DC rated voltage over ±800 kV is known as ultra-high voltage. UHV long distance and large capacity transmission can reduce line loss, thus achieving more economical trans-regional and trans-continental transmission, and a reliable, safe, economic, efficient and environmentally friendly power grid. Transmission power of 1000 kV AC can reach 4-5 GW, which is 4-5 times of 500 kV transmission power, and the theoretical line loss is only 1/4 of 500 kV.

December in 2007 witnessed construction of the world's highest DC voltage class Sichuan-Shanghai ±800 kV UHV transmission demonstration project.

1.4　Smart grid

American scientists proposed the new concept of "smart grid". Smart grid is the integration of information technology and power industry. In a smart grid, electricity is traded and used over the Internet, and electronic chips are embedded in every generation, transformation and end-use device and appliance. In the event of a failure, the power grid can heal itself, thus effectively preventing a blackout. The smart grid manages the demand side of the grid in a way that saves power even more.

A smart grid applies technologies, tools and techniques available now to bring knowledge to power-knowledge capable of making the grid more efficient. In the short term, a smart grid will function more efficiently, enabling it to deliver the level of service we've come to expect more affordably in an era of rising costs, while also offering considerable societal benefits — such as less impact on our environment. In the long run, the smart grid can be expected to spur the kind of transformation that the Internet has already brought to the way we live, work, play and learn.

1.5　Energy-saving technology of the system

Research to improve energy efficiency in the energy-intensive processes such as steel, aluminum, and cement ought to be focused more on the comprehensive system rather than individual equipment. For example, the improvement of single electric motor can raise efficiency by only a few percent, while redesigning the motor system can save 20%-60%. Speed regulation can save even more electricity. Combination of advanced sensors and control system helps to optimize the system.

1.6　Energy-saving technology of the process

High-efficiency process can produce high-quality new products and reduce waste generation. Its technologies include selective catalyst, advanced separation technology, measurement and control systems, new materials, advanced electric motor systems. In the long term, biotechnology and materials made from biomass appear particularly appealing.

Profound knowledge about chemistry, metallurgy and biotechnology could lead to significant advances in research and development of new manufacturing processes, including improved simulation, materials, instrumentation and intelligent systems. For example, the ongoing technologies of micro manufacturing systems (such as micro machines and micro chemical reactor), flexible processes, and in situ manufacturing are conducive to improving energy efficiency.

1.7　Superconducting technology

The research and application of superconducting technology will revolutionize the fields of electric power transmission, electric motors, transportation, computers and fundamental science experiments. The capacity of superconducting generator is 5-10 times larger than conventional generator, the volume is 1/2 smaller, and the efficiency is 50% higher. High-temperature superconductor can greatly reduce line loss and improve the reliability of the power system. In 2006, the bi-based high-temperature superconducting cable developed in China was successfully connected to the power grid. The 500 m high-temperature superconducting cable developed in Japan has been on trial.

By the middle of the 21st century, "superhighways" using high-temperature superconducting transmission lines, transformers and current controllers will dominate.

1.8　Nanotechnology

Nano-materials serve as the basis of nano-science and are related to energy. Carbon nanofibers and carbon nanotubes (CNTS) have been developed. Carbon nanotubes are ideal hydrogen storage materials, and can be used to develop more sensors and control systems with smaller volume and more reliable functions. It will contribute to advances in energy efficiency and low-carbon economic growth.

1.9　Energy storage

The existing energy storage technology mainly includes: pumped storage, compressed air storage and chemical battery. The technologies being developed are: energy storage technology concerning mechanism (flywheel, compressed air), electrochemistry (advanced battery, reversible fuel cell, hydrogen), electricity (supercapacitor, superconducting magnetic storage). The key to all the technical problems with energy storage is mainly cost reduction. The introduction of advanced

energy storage technology enables the entire system to improve energy efficiency and reduce carbon emissions.

The efficiency of a typical coal-fired power plant drops from 38% during the day to 28%-31% at night. Off-peak storage keeps the plant running at high efficiency, and a storage efficiency of more than 85% can reduce emissions of carbon dioxide.

2. Clean energy technology

To deliver reliable, emission-free and cost-competitive energy is what we strive for in developing clean energy. Developing and applying clean, new and renewable energy technology may be the least hope for the solution of severe energy shortage, environmental problems and so on.

2.1 Wind power generation

Since 2000, global wind power installed capacity has increased at an annual rate of more than 30%, reaching 74,223 MW in 2006. Wind power technology continues to improve and has become one of the most competitive renewable energy. By 2025, wind power in some areas could account for 10-20 percent of installed capacity. Wind turbine tends to become large-scale, with typical type of 1.5-3.0 MW wind turbine. Europe has mass production of 3.6 MW wind turbine. The United States has developed 7 MW wind turbines and the United Kingdom is developing 10 MW wind turbines. Wind power is moving offshore. At the end of 2007, a 100 MW wind power project was launched in the East China Sea.

Current focus of research and development regarding wind power technology is high tower, light blade, advanced blade design, direct drive system, variable-speed constant-frequency, intelligent control, and light and durable structural components.

2.2 Photovoltaic

In recent years, the global sales of photovoltaic cells have increased by more than 40% annually, reaching 1760 MW in 2005, and the cost of power generation has been reduced to 0.25 USD/kWh. Photovoltaic cells come to a new high tech industry.

Thin film photovoltaic cells (see Fig. 10.2) are being developed in 40 countries. CIS solar cell, with thinness of only 1/100 of silicon battery, can be covered on glass, plastic and other materials. It is soft, light, durable, and less costly because of being free from the use of the rare and expensive silicon. It can be applied in a variety of fields such as consumer electronics, remote monitoring, military, outdoor and indoor power supply. In 2006, the United States initiated the construction of a thin film photovoltaic cell plant with an annual capacity of 430 MW. In 2007, Sharp Electronics in Japan decided to build a thin film photovoltaic plant with an annual capacity of 1 GW.

Figure 10.2   Thin film photovoltaic cells integrated with the architecture

Source: http://www.sohu.com/a/24482499_114771.

2.3   Biomass energy conversion

Biomass energy can be converted into electricity, fuel and other products. In the coming decades, molecular genetics will be employed to improve the economics of all forms of biomass energy, developing technologies that combine some biomass to produce ethanol, electricity and chemicals. Improvements in fuel cells will increase the demand for and value of biogas.

2.4   Hydroelectricity

In 2006, the world's hydropower generating capacity was 3,096.5 TWh, equivalent to 21.5% of the world's technically exploitable hydropower resources.

Advanced hydropower technology can eliminate the adverse effects of hydropower on the environment and increase the generation of electricity. It is necessary to investigate into the impact of hydropower projects on fish, renovation and life extension of the existing hydropower stations. Research needs to be done into the energy storage technology of hydropower as an intermittent renewable energy. Hydropower can improve power supply reliability of wind power and photovoltaic power generation.

2.5 Geothermal energy

There are four types of geothermal energy: (1) geothermal water or steam, buried in the depth of 100-4,500 m and the temperature of 90-350 ℃; (2) ground pressure geothermal energy, buried in a 3-6 km deep thermal storage layer at temperatures between 90 ℃ and 200 ℃; (3) geothermal energy from dry hot rocks, which contains little or no water can reach temperatures exceeding 200 ℃; (4) magmatic geothermal energy, stored in molten magma at 700-1,200 ℃, which is the most abundant geothermal energy, and the depth to explore is about 3,000-10,000 m.

Up to now, only hot-water geothermal energy has been developed commercially. The geothermal energy below 150 ℃ can be directly utilized, and the geothermal energy exceeding 150 ℃ is used for power generation, and the installed capacity of the world geothermal power generation is 9,600 MW in 2006.

It is necessary to develop new technologies such as high temperature hard rock drilling, thermal reservoir prediction, fracturing and so on.

2.6 Ocean energy

Ocean energy mainly includes tidal energy, wave energy, tidal current energy, ocean temperature difference energy and ocean salinity difference energy. The ocean energy reserves are huge, but the energy density is low, and changes with time and space, with harsh development environment and enormous investment. So far only tidal power and small wave power generation units have been commercialized. The tidal current energy, temperature difference energy and salinity difference energy are all in the experimental stage.

2.7 Nuclear energy

In recent years, concerns over oil prices, energy security and increased carbon dioxide emissions have revived nuclear power.

The research and development priorities are given to advanced pressurized water reactor and high temperature gas-cooled reactor, the design of new high efficiency reactor, the non-proliferation reactor and fuel technology, the design and application of advanced low power reactor, and advanced nuclear fuel and waste storage technology, breeder reactor and nuclear fusion technology.

3. Carbon capture and storage

For the sake of environmental protection, measures should be taken to capture and store the enormous carbon dioxide released during energy production. It is critical to reduce carbon emissions. There are six methods for carbon dioxide capture and storage: (1) separated and captured from energy systems; (2) stored in the ocean; (3) stored in terrestrial ecosystems, such as forests, plants, soil, crops, pastures, tundra, wetlands; (4) sealed in the strata; (5) advanced biological technology; (6) advanced chemical methods.

Carbon dioxide capture technology in power plants includes solvent separation, low temperature treatment, gas separation membrane, adsorption and so on. Capture of carbon dioxide would consume energy and could add 2 cents per kWh to the cost of electricity generation. China has launched a demonstration project to capture carbon dioxide from power plants.

The oceans are the largest potential reservoir for store of carbon dioxide. Due to weak interaction between the surface and deep ocean, it takes a very long time for carbon dioxide to return to the atmosphere after injected into the deep ocean. In Norway, oil companies have developed technology to pump carbon dioxide into the deep sea, and inject the 2,800 tons of carbon dioxide they have recovered each year into seabed sandstone at depths of 800 meters. However, little is known about the effects of such methods on the marine environment, such as changes in the pH-value of seawater.

Depleted oil and gas reservoirs, deep saltwater reservoirs are suitable for carbon dioxide storage and are more economical. That's because less is spent on exploration of carbon dioxide storage, and even some equipment for oil and gas production can be used to inject carbon dioxide. The IEA (International Energy Agency) estimates that the world's depleted oil and gas reservoirs could store 45 percent of global carbon dioxide emissions, while deep saltwater reservoirs could store more than 400bn tonnes. Injection of carbon dioxide into the reservoir is also an important enhanced oil recovery (EOR) technique.

New words and phrases

adsorption *n.* 吸附（作用）

alloy steel 合金钢

breeder reactor 快中子增殖反应堆

carbon capture and storage 碳捕捉和封存

depleted *adj.* 耗尽的；废弃的

deployment *n.* 调度，部署

EOR（enhanced oil recovery） *abbr.* 强化采油

ferritic *adj.* 铁素体的；铁氧体的

fracturing *n.* 压裂

gas separation membrane 气体分离膜

high-temperature superconductor 高温超导材料

in-situ manufacturing 现场生产

line loss 线路消耗

magmatic *adj.* 岩浆的

metallurgy *n.* 冶金

molecular genetics 分子遗传学

nanofiber *n.* 纳米纤维

non-proliferation reactor 防扩散反应堆

nuclear fusion technology 核聚变技术

polymer *n.* 聚合物

salinity *n.* 盐度；盐分

selective catalyst 选择性催化剂

solvent separation 溶剂分离

spur *v.* 激励；促进

strata　*n.*　地层

supercapacitor　*n.*　超级电容器

superconducting technology　超导技术

tertiary recovery technology　强化采油技术

thin film photovoltaic cell　薄膜光伏电池

transmission power　传输功率

tundra　*n.*　冻原;冻土地带

ubiquitous　*adj.*　普遍存在的;无所不在的

variable-speed and constant-frequency　变速恒频

 Exercises

Ⅰ. Warming-up questions

Directions: Give brief answers to the following questions.

1. Why do we need to initiate energy technological innovation?

2. What potential disadvantage do you know about UHV transmission?

3. What technologies have been adopted in development of fossil fuels?

Ⅱ. Technical terms

Directions: Please give the Chinese or English equivalents of the following terms.

1. ultra-high voltage transmission　　2. nanotechnology

3. EOR (enhanced oil recovery)　　4. superconducting technology

5. line loss　　6. hot dry rock

7. 高温超导材料　　8. 超临界火电机组

9. 碳收集和封存　　10. 核聚变技术

11. 薄膜光伏电池　　12. 生物质能发电

Ⅲ. Blank filling

Directions: Complete the following sentences by choosing words or phrases given below.

| diverse | residential | transition | sphere | deployment |

1. Hydrogen and fuel cell technologies offer greater personal choice in the _____ to a low-carbon economy, given their similar performance, operation and consumer experience to fossil-fuelled technologies.

2. Advancing the low-carbon technology revolution will involve millions of choices by a myriad of stakeholders — all individuals acting in personal or professional _____.

3. The suitability of hydrogen and fuel cells varies between transport modes and reflects the _____ nature of the transport sector, which spans land, sea and air, plus freight and passengers.

4. Fuel cell buses have seen substantial early _____, with 7 million kilometres of operational experience so far in Europe.

5. Heat and hot water account for 60%-80% of final energy consumption in _____ and commercial buildings across Europe. Emissions from heating need to be reduced rapidly and largely by 2050.

Ⅳ. Translation

Directions: Translate the following passages from English into Chinese or from Chinese into English.

Passage 1

This paper provides a comprehensive state-of-the-art update on hydrogen and fuel cells across transport, heat, industry, electricity generation and storage, spanning the technologies, economics, infrastructure requirements and government policies. It defines the many roles that these technologies can play in the near future, as a flexible and versatile complement to electricity, and in offering end-users more choice over how to decarbonise the energy services they rely on.

Passage 2

各国为解决能源危机,在积极制定节能措施、提高能源利用率、降低能源消耗政策的同时,开始逐步向多元能源结构过渡。一方面积极开发新的高效节能技术,另一方面大力开发新能源的利用技术,主要是核能技术,太阳能利用技术,地热能、风能、生物能及海洋能利用技术等。

 Additional reading

Keys to Exercises

Unit 1

I. Warming-up questions
略。

II. Technical terms
1. 臭氧层空洞
2. 电磁辐射
3. 二次能源
4. 水力发电
5. 潮汐能
6. 存储容量
7. traditional energy
8. renewable energy
9. energy conversion
10. internal combustion engine
11. mechanical energy
12. fossil fuel

III. Blank filling
1. convertible
2. degradation
3. radiant
4. electromagnetic
5. dissipate

IV. Translation

Passage 1

按照既定政策,到 2040 年能源需求将以每年 1% 的速度增长。以太阳能光伏发电（PV）为首的低碳能源占了这一增长的一半以上,而由液化天然气（LNG）贸易增长推动的天然气占了三分之一。到本世纪 30 年代,石油需求将趋于平稳,煤炭使用量将有所下降。以电力为主导的能源部的一些部门正在经历迅速的转变。一些国家,尤其是那些有"零净"愿望的国家,在改变其供应和消费的所有方面都走得很远。然而,清洁能源技术的发展不足以抵消全球经济扩张和人口增长带来的影响。排放量的增长速度有所放缓,但由于在 2040 年之前还没有达到峰值,全球距离共同的可持续发展目标还有很长的路要走。

Passage 2

Nuclear energy is technically mature and can be developed and utilized on a large scale. It is an energy source to make up for the shortage of fossil fuels, reduce environmental pollution and realize large-scale industrial application. The proven nuclear fission fuels on the earth, uranium and thorium, are more than 20 times the total energy of fossil fuels in terms of their energy content. In addition, there is still a large amount of deuterium on the earth as fuel for nuclear fusion. If commercial controlled nuclear fusion is achieved, the reserves of deuterium and tritium on the earth can be used for 10 billion years at the current level of world energy consumption.

Unit 2

I. Warming-up questions

略。

II. Technical terms

1. 脱硫作用
2. 锅炉给水
3. 冲动式汽轮机
4. 电势
5. 排出的气体
6. 交流电
7. thermal power plant
8. working medium

9. direct current   10. slag discharge

11. suspension combustion boiler   12. condensing turbine

III. Blank filling

1. dedust

2. regenerative

3. induce

4. fluctuations

5. thermal efficiency

IV. Translation

Passage 1

在有机朗肯循环（ORC）系统中,发电机直接连接到为高速驱动而设计的涡轮膨胀机上,从而减小了尺寸并提高了效率。此外,发电机转子将根据ORC系统的运行条件以可变速度运行。但是,由于ORC系统的不可预测性导致涡轮转速的波动或起伏,发电机暴露在转速波动中,这又会引起明显的振动和噪声。

Passage 2

Boiler is a kind of energy conversion equipment. It is a device that uses the thermal energy released by the combustion of fuel or other thermal energy to heat working fluid or other fluids to certain parameters. The original meaning of the pot refers to a water container heated on the fire, the furnace refers to the place where the fuel is burned, and the boiler includes the pot and the furnace. The hot water or steam generated in the boiler can directly provide the required thermal energy for industrial production and people's lives, and can also be converted into mechanical energy through steam power plants, or into electrical energy through generators. Boilers that provide hot water are called hot water boilers, which are mainly used in daily life. There are also a small number of applications in industrial production. The boiler that generates steam is called a steam boiler, often referred to as a boiler, and is mostly used in thermal power stations, ships, locomotives, and industrial and mining enterprises.

Unit 3

Ⅰ. Warming-up questions
略。

Ⅱ. Technical terms

1. 风力发电
2. 海上风电场
3. 旋转叶片
4. 纵轴
5. 容比
6. 无功功率
7. aerodynamics
8. incandescent light bulb
9. doubly fed machine
10. rotor hub
11. operating reserve
12. pumped-storage plant

Ⅲ. Blank filling

1. intermittent
2. variable
3. penetration
4. specification
5. address

Ⅳ. Translation

Passage 1

风力发电场由许多单独的风力涡轮机组成,这些涡轮机连接到电力传输网络。陆上风能是一种廉价的电力来源,与煤炭或天然气发电厂竞争,也许在许多地方比煤炭或天然气发电厂还便宜。陆上风力发电场对景观也有影响,因为它们通常需要分布在比其他发电站更广阔的土地上,需要在野外和农村地区建设,这可能导致"农村工业化"和栖息地的丧失。海上风电比陆地风电更稳定、更强,海上风力发电场的视觉冲击较小,但建设和维护成本较高。小型陆上风力发电场可以向电网提供一些能量,也可以向隔离的离网位置提供电力。

Passage 2

Wind is an intermittent energy source, which cannot make electricity nor be dispatched on demand. It also gives variable power, which is consistent from year to year but varies greatly over shorter time scales. Therefore, it must be used together with other electric power sources or storage to give a reliable supply. As the proportion of wind power in a region increases, more conventional power sources are needed to back it up (such as fossil fuel power and nuclear power), and the grid may need to be upgraded. Power-management techniques such as having dispatchable power sources, enough hydroelectric power, excess capacity, geographically distributed turbines, exporting and importing power to neighboring areas, energy storage, or reducing demand when wind production is low, can in many cases overcome these problems. Weather forecasting permits the electric-power network to be readied for the predictable variations in production in case that occur.

Unit 4

I. Warming-up questions
略。

II. Technical terms
1. 太阳能
2. 光伏
3. 集中式太阳能发电
4. 电流
5. 可再生能源
6. 太阳能电池板
7. photovoltaic power station
8. grid energy storage
9. thermoelectric system
10. landfill gas
11. thermal energy storage
12. artificial photosynthesis

III. Blank filling
1. conversion
2. renewable
3. installation

4. reflector

5. capacity

IV. Translation

Passage 1

太阳能发电是从太阳光到电力的能量转换——无论是直接使用光伏 (PV)，还是间接使用集中式太阳能发电，或它们的组合。集中式太阳能发电系统使用透镜或镜子和太阳跟踪系统将大面积的阳光聚焦为一小束光。光伏电池利用光伏效应将光转换为电流。

Passage 2

The array of a photovoltaic power system, or PV system, produces direct current (DC) power which fluctuates with the sunlight's intensity. For practical use this usually requires conversion to certain desired voltages or alternating current (AC), through the use of inverters. Multiple solar cells are connected inside modules. Modules are wired together to form arrays, then tied to an inverter, which produces power at the desired voltage, and for AC, the desired frequency / phase.

Unit 5

I. Warming-up questions

略。

II. Technical terms

1. 静电

2. 热电效应

3. 摩擦起电

4. 电路

5. 玻璃电荷

6. 树脂电

7. photoelectric effect

8. luminiferous ether

9. galvanometer

10. diamagnetic

11. special relativity

12. quantum electromagnetic theory

III. Blank filling

1. friction

2. charge

3. conductive

4. unified

5. permeability

IV. Translation

Passage 1

根据麦克斯韦方程,真空中的光速是一个常数,它只依赖于自由空间的电介常数和磁导率。这违反了经典力学的一个长期基石——伽利略不变性。调和这两种理论(电磁学和古典力学)的一种方法是假设光通过一个有光以太传播。然而,随后的实验未能检测到乙醚的存在。1905年,在亨德里克·洛伦兹和亨利·庞加莱的重要贡献之后,阿尔伯特·爱因斯坦用狭义相对论解决了这个问题,用一种与经典电磁学兼容的运动学新理论取代了经典运动学。

Passage 2

Electricity, as a branch of classical physics, is well developed in terms of its fundamental principles, and can be used to explain electromagnetic phenomena in the macroscopic field. In the 20th century, with the development of atomic physics, atomic nucleus physics and particle physics, human's understanding went deep into the micro field, in the interaction between charged particles and electromagnetic fields, the classical electromagnetic theory encountered difficulties. Although the classical theory has given some useful results, many phenomena cannot be explained by the classical theory. The limitation of the classical theory is that the description of charged particles ignores their volatility and the description of electromagnetic waves ignores their particle nature. According to quantum physics, both matter particles and electromagnetic fields are both particle and volatile. Under the promotion of microscopic physics research, classical electromagnetic theory developed into quantum electromagnetic theory.

Unit 6

I. Warming-up questions

略。

II. Technical terms

1. 电力电子
2. 变压器
3. 发电厂
4. 水电站
5. 配电站
6. 原动机
7. three-phase
8. steam turbine
9. voltage drop
10. input voltage
11. generator
12. fault clearance

III. Blank filling

1. utilization

2. substation

3. transmission

4. converting

5. synchonorous

IV. Translation

Passage 1

热力学循环由一系列热力学过程组成,包括热量的传递、功的输入和输出,同时改变系统内的压力、温度和其他状态变量,最终使系统恢复到初始状态。一方面,在热力学循环的过程中,工作流体(系统)可以将热量转换为有用功,并将剩余的热量释放到冷却散热器中,从而充当热力发动机。另一方面,该循环也可以反过来,利用功将热量从冷源转移并将其传递到较热的散热器,从而充当热泵。在循环的每个点,系统都处于热力学平衡状态,因此循环是可逆的。

Passage 2

Rapid deployment of solar photovoltaic, led by China and India, helps solar become the largest source of low-carbon capacity by 2040, by which time the share of all renewables in total power generation reaches 40%. In the European Union, renewables account for 80% of new capacity and wind power becomes the leading source of electricity soon after 2030, due to strong growth both onshore and offshore. Policies continue to support renewable electricity worldwide, increasingly through competitive auctions rather than feed-in tariffs, and the transformation of the power sector is amplified by millions of households, communities and businesses investing directly in distributed solar PV. Electricity is the rising force among worldwide end-uses of energy, making up 40% of the rise in final consumption to 2040 — the same share of growth that oil took for the last twenty-five years.

Unit 7

I. Warming-up questions
略。

II. Technical terms

1. 人工呼吸
2. 一次设备（主设备）
3. 功率损耗
4. 电量
5. 隔离开关
6. 带电体
7. electric shock
8. relay protection
9. insulated gloves
10. chest compression
11. electrothermal burn
12. excitation current

III. Blank filling

1. insulate
2. diffuse
3. electrify
4. static
5. step-up

IV. Translation

Passage 1

所有断路器系统在操作上都具有共同的特征,但是具体细节会因断路器的电压等级、额定电流和类型而有很大不同。断路器必须首先检测故障状况。在小型市电和低压断路器中,这通常是在设备内部完成的。通常,利用电流加热或磁效应进行。用于大电流或高电压的断路器通常装有保护继电器先导装置,以检测故障情况并操作开放装置。尽管某些高压断路器是独立的,并配有电流互感器、保护继电器和内部控制电源,但这些设备通常需要单独的电源,比如电池。

Passage 2

Electricity can bring disasters as well as benefits to mankind. Statistics show that electrical accidents account for a large proportion of industrial accidents. Taking the death toll in construction work as an example, in 2011, 7. 34% of all construction accidents in China were caused by electric shock. For every 150 million kWh of electricity used in China, 1 person died from electric shock, while only 1 person died from electric shock in the US, Japan and other countries for every 2-4 billion kWh of electricity used. According to statistics, electrical fires account for about 20% of all fires, causing huge casualties and economic losses. For example, of more than 5,000 fires that occurred in Beijing last year, electrical fires ranked first and have become the biggest fire hazard.

Unit 8

I. Warming-up questions

略。

II. Technical terms

1. 最大负荷
2. 插电式混合动力汽车
3. 超导电性
4. 数据采集
5. 负载能力
6. 参数测量
7. power density
8. transmission capacity

9. fuel cell

10. onboard battery

11. electric meter

12. depletion of energy

III. Blank filling

1. aggregate

2. seamless

3. bidirectional

4. depletion

5. disperse

IV. Translation

Passage 1

可再生能源中,智能电网是联通可再生能源生产与电网的部门或通信区域。但是,可再生能源生产与智能电网之间的联通带来了许多挑战,例如稳定性问题、复杂的操作过程和远程控制。电网不仅是从发电厂到变电站的电力传输系统,而且还覆盖了从变电站到个人用户的电力分配。智能电网系统将包含许多具有挑战性的过程和技术,例如监视和分析,自动化或控制(高压设备的主动控制、鲁棒性、可靠性、安全性和效率等),分布式能源如微电网、可再生能源、固体氧化物、燃料电池、电池存储系统的集成和控制等。

Passage 2

The research and development of smart grid covers almost all the emerging industries that are currently booming, including high-tech fields such as computers, new energy, electronic information, communication technology, system control technology and so on. It is conceivable that in the future, the smart grid may have an unpredictable far-reaching impact on various industries like the rise of the Internet, and may even trigger a new round of industrial transformation in the energy system. In the future, the smart grid will continue to make technological breakthroughs and structural innovations in the direction of unmanned, information and intelligent, and the scope covered by the smart grid will also gradually expand.

Unit 9

Ⅰ. Warming-up questions

略。

Ⅱ. Technical terms

1. 能效
2. 油田采收率
3. 可再生能源配额制
4. 逐步淘汰
5. 电动汽车
6. 海上风电
7. energy saving and emission reduction
8. environmentally friendly society
9. eliminate backward capacity
10. energy-saving renovation
11. high-end equipment
12. strategic emerging industry

Ⅲ. Blank filling

1. availability
2. curb
3. phasing out
4. concerted
5. incompatible with

Ⅳ. Translation

Passage 1

上周在达沃斯论坛上,人们对气候变化问题的紧迫性和关注度显著提高,这一点是大家喜闻乐见的,但要将雄心转化为行动,则需企业及社会活动家们一致支持政策变革,关注排放增长的地区和能源部门,并就能源产业在解决能源脱碳问题所起作用方面达成共识。

Passage 2

International cooperation in energy is based on frequent and profound exchange among exporters, consumers and transferred countries. The international community should further strengthen bilateral and multilateral relations, launching dialogues on energy efficiency, energy conservation and environmental protection, energy management and energy policy. Improve the monitoring and emergency response mechanism of the international energy market. Increase interaction in the exchange of information, personnel training and coordination of actions.

Unit 10

I. Warming-up questions
略。

II. Technical terms
1. 特高压输电

2. 纳米技术

3. 强化采油

4. 超导技术

5. 线路消耗

6. 干热岩

7. high-temperature superconductor

8. ultra supercritical unit

9. carbon capture and storage

10. nuclear fusion technology

11. thin film photovoltaic cell

12. biomass energy generation

III. Blank filling
1. transition

2. spheres

3. diverse

4. deployment

5. residential

IV. Translation
Passage 1

本文介绍氢气和燃料电池在交通、热能、工业、发电和储存方面的最新进展,涵盖技术、经济、基础设施要求和政府政策。作为对电力的灵活多样的补充,这些技术在不久的将来会发挥重大作用,终端用户将有更多的选择去决定如何在他们赖以生存的能源服务中实现脱碳。

Passage 2

Harassed by energy crisis, countries engage in energy conservation measures and policies for higher energy efficiency and lower consumption. Meanwhile, they gradually head for a diversified energy structure. Much effort has been put into advanced energy-saving technologies. Simultaneously, research is underway about the technologies to deploy new energy, mainly technologies about nuclear energy, solar energy, geothermal energy, wind energy, bio-energy, and ocean energy.

References

[1]　陈砺,严宗诚,方利国.能源概论 [M]. 2 版.北京：化学工业出版社,2019.

[2]　陈卫东.智能电网下新型智能电表的功能、应用及特点.科技与企业 [J].
2013 (24).

[3]　郭莉鸿.安全用电 [M]. 北京：中国电力出版社,2007.

[4]　国网能源研究院有限公司.中国能源电力发展展望 2017 [M]. 北京：中国
电力出版社,2017.

[5]　刘洪恩.新能源概论 [M]. 北京：化学工业出版社,2018.

[6]　梁绍荣,刘昌年,盛正华.普通物理学 [M]. 3 版.北京：高等教育出版社,
2005.

[7]　王长贵,王斯成.太阳能光伏发电实用技术 [M]. 2 版.北京：化学工业出
版社,2009.

[8]　王革华.新能源概论 [M]. 北京：化学工业出版社,2006.

[9]　徐立娟.电力电子技术 [M]. 北京：人民邮电出版社,2010.

[10]　易跃春.风力发电现状、发展前景及市场分析 [J]. 国际电力,2004,
8 (5).

[11]　赵明学,周英莉,杨晓华,等.能源英语 1 [M]. 北京：知识产权出版社,
2016.

[12]　赵明学,周英莉,杨晓华,等.能源英语 2 [M]. 北京：知识产权出版社,
2016.

[13]　赵艳玲,张群,张仕海.电工电子技术 [M]. 成都：西南交通大学出版社,
2015.

［14］ 赵子涵. 智能电网的发展概述与展望. 中国科技纵横［J］. 2017（24）：128-129.

［15］ 张文木. 中国能源安全与政策选择. 世界经济与政治［J］. 2003,5（3）：11-16.

［16］ 国际能源署. 世界能源展望2004［M］. 朱起煌,等译. 北京：中国石化出版社,2006.

［17］ 朱永强,尹忠东. 电力专业英语阅读与翻译［M］. 北京：中国水利水电出版社,2017.

［18］ S. R. Wenham, M. A. Green, M. E. Watt,等. 应用光伏学［M］. 狄大卫,高兆利,施正荣,等译. 上海：上海交通大学出版社,2008.

［19］ 严陆光,夏训诚,周凤起,等. 我国大规模可再生能源基地与技术的发展研究［J］. 电工电能新技术,2007,26（1）：13-24.

［20］ Dileep G. A Survey on Smart Grid Technologies and Applications［J］. Renewable Energy,2020（146）：2589-2625.

［21］ Hyung-Seok Park, Hong-Jun Heo, Bum-Seog Choi, et al. Speed Control for Turbine-Generator of ORC Power Generation System and Experimental Implementation［J］. Energies,2019,12（2）：1-13.

［22］ Paul Gipe. Wind Power：Renewable Energy for Home, Farm, and Business［M］. James & James（Science Publishers）,2004.

［23］ M. S. Hossain, N. A. Madlool, N. A. Rahim, et al. Role of Smart Grid in Renewable Energy：An Overview［J］. Renewable and Sustainable Energy Reviews,2016（60）：1168-1184.

［24］ Said Al-Hallaj. More than Enviro-friendly, Power & Energy［J］. IEEE,2004, 2（3）：16-22.

［25］ Vlatko Vlatkovic. Alternative Energy：State of the Art and Implications on Power Electronics［C］. Proc IEEE APEC,2004.

［26］ X. S. Cai. Renewable Energies,Present & Future［J］. 电工电能新技术, 2005,24（1）：69-75.

与本书配套的二维码资源使用说明

  本书部分课程及与纸质教材配套数字资源以二维码链接的形式呈现。利用手机微信扫码成功后提示微信登录，授权后进入注册页面，填写注册信息。按照提示输入手机号码，点击获取手机验证码，稍等片刻收到 4 位数的验证码短信，在提示位置输入验证码成功，再设置密码，选择相应专业，点击"立即注册"，注册成功。(若手机已经注册，则在"注册"页面底部选择"已有账号？立即注册"，进入"账号绑定"页面，直接输入手机号和密码登录。) 接着提示输入学习码，需刮开教材封面防伪涂层，输入 13 位学习码(正版图书拥有的一次性使用学习码)，输入正确后提示绑定成功，即可查看二维码数字资源。手机第一次登录查看资源成功以后，再次使用二维码资源时，只需在微信端扫码即可登录进入查看。